FOREWORD

This publication, the latest in a series designed to provide students and others with a front-row seat at the frontiers of biomedical research, coincides with one of the most important advances in the history of science. Early in 2001, two teams of researchers published the complete genetic readout of humans. They listed all of the chemical letters that spell out the instruction manual for our species. Observers immediately hailed the publication of the human genome for both its value to scientists and its potential to speed new medical applications.

Yet the impact of the human genome project on humanity cannot be described in practical terms alone. As this publication, *The Genes We Share*, makes clear, one of the most wondrous revelations of this new field of genomics is that we humans have a genetic plan that is remarkably similar to that of other species. We are much less unique than we once believed.

Such a revelation reaches beyond science to our concept of who we are. Our sense of identity may soon be changed in the same way that the Renaissance discovery about the Earth revolving around the Sun transformed mankind's view of its place in the Universe. Gradually, we humans came to understand that, instead of being at the very center of existence, we live on a planet much like other planets, derived from the same burst of creation. So it is with these new genetic discoveries. We are just beginning to realize that many of the genes that guide our lives are largely the same as those in other creatures, even such simple organisms as worms or yeast. Our biological software is more complex, but constructed from a common genetic kit. Like our planet, we are part of a larger whole, ensemble players working from a shared evolutionary script.

Researchers are now routinely logging onto the Internet to check whether a genetic sequence has been discovered in other organisms and, if so, what is known about it. Beyond the research laboratory, physicians are using genetic tests to determine whether patients are at greater risk for getting certain cancers or other diseases. New "biochips" will improve both diagnosis and treatment for patients whose illnesses appear identical but differ at the molecular level. Many other innovations are under development, or will be soon.

Often, these advances are accompanied by difficult social and ethical questions. Should parents test to see whether their unborn child has a genetic defect, and how should they respond if it does? Should an insurance company, an employer, or *anyone* besides a physician and patient be allowed to see the person's genetic profile?

The new genetic information that is changing our world results directly from our society's patient investment in basic science. Researchers have devoted lifetimes to studying the genes of yeast and flies. Their work may have seemed pointless or obscure to nonscientists, but now everyone can see the payoff. The genetic knowledge that scientists developed so painstakingly about small organisms turns out to be directly applicable to humans.

Basic biomedical science has taken off like a rocket over the past half-century and now, as it converges with equally remarkable advances in computer science, it is providing us with both profound insights about life and the promise of improved medical care. Our capacity to see the world with scientific eyes—to ask questions and evaluate evidence, to make hypotheses and test conclusions—sets us apart from the yeast, worms, flies, and mice whose genes we share. We have the wisdom to understand that the discoveries described in this publication are not a challenge to our humanity but, instead, an opportunity to celebrate our place on Earth with a deeper appreciation of who we really are.

Thomas R. Cech

Thomas R. Cech, Ph.D.
President
Howard Hughes Medical Institute

Laboratories of Yigong Shi, Princeton University, and Xiaodong Wang, U.T. Southwestern Medical Center

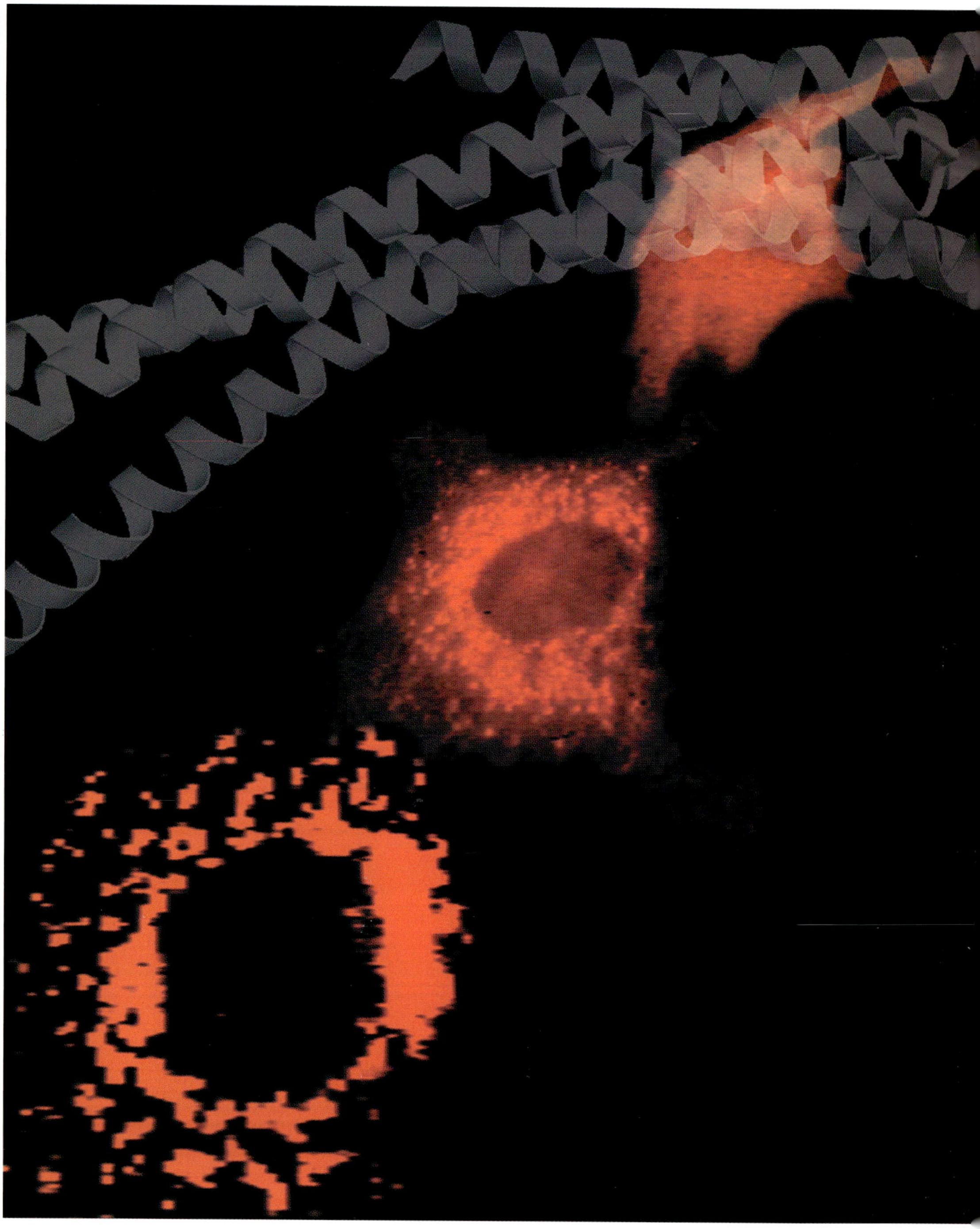

The signals that tell some animal cells they must commit suicide for the good of the whole organism turn out to be very similar in worms, flies, and humans. Here a signaling protein called Smac/DIABLO (red) is shown moving out of its usual place in a mitochondrion (lower left) to help activate a cell's death program. New drugs mimicking part of this protein may help kill cancer cells.

A REPORT FROM THE HOWARD HUGHES MEDICAL INSTITUTE

THE GENES WE SHARE WITH YEAST, FLIES, WORMS, AND MICE

New Clues to Human Health and Disease

Introduction

It's fortunate that nature keeps reusing the same building blocks while constructing organisms as different as yeast cells, worms, flies, mice, and humans. This frugality makes it possible for us to learn what our own genes do (which is still largely a mystery) by analyzing similar genes in tiny, rapidly reproducing "model organisms"—organisms that scientists can manipulate with ease in the lab and have come to know very well.

Our genes are actually so similar to those of yeast, fruit flies, nematode worms, and mice that in some cases human genes can be substituted for theirs and work just as well. However, even a slight similarity between a human gene and one from a model organism is generally enough to give scientists valuable hints about the human gene's role in health and disease.

Even though researchers are rightly proud of their tremendous achievement in sequencing the entire human genome (that is, deciphering the sequence of all 3 billion base pairs in human DNA), they would be unable to make much use of this treasure if they did not have access to the information provided by model organisms. This information is also vital to scientists' hopes of developing a new generation of tailor-made treatments for diseases ranging from Alzheimer to many types of cancer.

Key findings about human genes have come from studying the humble, blob-like cells of baker's or brewer's yeast, which one researcher calls, shockingly, our relatives. In 1996, yeast became the first eukaryote (an organism whose genetic material is enclosed in a cell nucleus) to have its entire genome sequenced. Ever since, it has remained at the forefront of research on genetics. Almost everything we know about the cell-division cycle, for instance, comes from experiments with yeast, and many new methods of analyzing genes were first tried out in yeast.

A good example is the do-it-yourself "microarrayer," a robot designed by HHMI investigator Patrick O. Brown at Stanford University. The robot can analyze all 6,000 genes in a yeast cell simultaneously and compare the activity of different sets of genes under various conditions. Whole-genome analysis of this sort promises a revolution in biology. To make the robot widely available, Brown has posted instructions for building it on the Web. Microarrayers have now spread to many labs around the world, where they are used to study not only yeast but other model organisms and even human cells.

Apart from yeast, which is unicellular, most model organisms consist of millions or even trillions of cells. They have brains, guts, and more complex behavior, which implies that their cells need to communicate with one another. The cells also need to act unselfishly, for the good of the whole organism, as do human cells.

The microscopic worm *Caenorhabditis elegans,* for instance, gave scientists the opportunity to analyze the mechanisms of cell suicide, or apoptosis, a startling phenomenon that is essential to animal health. Cell suicide has also been studied in flies and humans.

The fruit fly *Drosophila,* which behaves surprisingly like humans when it gets drunk, turns out to be a good source of information about the activity of brain neurons and the genetics of addiction, as well as about cancer.

And the mouse, like all mammals, is genetically so close to humans that segments of its genome are sometimes hard to tell apart from ours. The mouse model will be invaluable in medical trials of experimental drugs.

Each model organism has its own set of advantages and disadvantages for researchers, who must carefully choose which model is best for a particular experiment. Increasingly, though, scientists can shuttle between species—including some new model organisms such as fish—to make the most of the special qualities of each one.

Despite all the obvious differences between, say, a fly and a person, and despite wide variations in the number of genes and chromosomes they contain, each organism has approximately the same number of core proteins.

"Complexity does not come from the number of genes but from the way in which they are used," explains Gerald M. Rubin, HHMI vice president for medical research, who heads the Berkeley *Drosophila* genome project. "Humans may have four copies of a gene where the fly has one, but if you look at the core proteome—the core set of parts—they're not that different."

Recently scientists examined a set of 289 genes that are known to be mutated or deleted in various human diseases. When they looked for similar genes in model organisms, they found a stunning number of such genes in the fruit fly: 177, or 61 percent of the total. When they limited their search to genes that are mutated in human cancers, they found an even higher proportion (68 percent) of parallel genes in the fruit fly. Worms and yeast also turned out to have large numbers of genes that resemble human disease genes (see the poster at the end of this report).

Researchers hope they will soon learn enough about such genes and the proteins they encode to be able to short-circuit some of the faulty protein interactions that cause disease. They also hope their findings will bring new levels of precision to human physiology and medicine.

If they succeed at both goals, as seems increasingly likely, it will be thanks to the genes we share with yeast, flies, worms, and mice—and to the towering evidence of the unity of life.

Maya Pines, *Editor*

A ROBOT THAT TRACKS *ALL* the Genes in a Cell REVEALS KEY PATTERNS

DISCOVERING WHAT THE GENES IN A LIVING CELL DO—STARTING WITH YEAST

By Maya Pines

At 7 a.m., Patrick Brown strides into his lab at the Stanford University Medical School, relatively late because he had to drop off his twins at school. His research team is already engrossed in the day's activities. A young woman hurries by, carrying a tray of flasks. "This new sequence we got—wow!" she calls out exuberantly as she passes him. Brown smiles. He moves on to his small, cramped office. At once a young man pokes his head in to say, "The work is going real well."

Brown and his team are on a roll—and they know it. Every day they generate more ideas and projects than they can possibly carry out. They also produce far more data than they have time to analyze. Tall, thin, and intense, Pat Brown simply cannot speak slowly—his words tumble out in quick spurts as he describes his work. Nor can he hide his excitement about the speed with which his kind of research—whole-genome analysis, in which the activity of *all* the genes in a cell is examined at once—is making progress and sweeping through biology.

It's all part of the astounding growth in knowledge about the genetic instructions inside living creatures, a growth that has caught even the genetics community by surprise. So many genes are being uncovered so rapidly, and in so many different organisms, that scientists cannot keep up with the flood. Studying these genes one by one seems far too slow. To move things along, several researchers, including Brown, an HHMI investigator, have devised ingenious ways of tracking the role of every gene in the genome simultaneously.

This global approach enables scientists to see changing patterns of activity in a cell's genes when the cell is placed in different conditions and to follow such changes over time. Then the researchers try to figure out the patterns' meaning: Why is one group of genes in the cell increasing its activity just when another cluster is shutting down, for instance? Why are so many genes active during a process

Patrick Brown peers at his microarrayer, a robot that can record the activity of thousands of genes simultaneously.

that was supposed to require only one gene? How does the activity pattern of genes in one type of cell differ from their pattern in another type?

Such questions could never have been asked before, and scientists are rushing to provide the answers. "It's breathtaking to see all this progress," says Ira Herskowitz, a pioneering yeast researcher who collaborates with Brown on some experiments. He predicts that whole-genome analysis will reveal the functions of thousands of newly identified genes. It will identify genes that regulate key pathways in a cell. It will enable physicians to make far more specific diagnoses. And it will lead to individually tailored therapies.

No wonder more and more scientists around the world want to do whole-genome research. But there is a problem: Until now, few researchers have had access to the whole genome of any living organism.

Brown belongs to the privileged group that deals with entire genomes because he chose to work on the most thoroughly analyzed of the several model organisms that geneticists use for experiments. Bypassing the small gray laboratory mouse (*Mus musculus*), the tiny fruit fly (*Drosophila melanogaster*), and the even tinier roundworm (*Caenorhabditis elegans*), Brown zeroed in on baker's yeast (*Saccharomyces cerevisiae*), also known as brewer's yeast, a microscopic, single-celled organism that looks like a blob.

Biologists enjoy working with yeast because it is almost as tractable as the bacteria and viruses that researchers used to devise recombinant DNA technology in the 1970s yet has many of the characteristics of mammalian cells. Everything in yeast's environment can be controlled with ease. At the same time, yeast genes can be manipulated in any way a researcher desires. Genes can be knocked out, added to specific chromosomes, replaced with other genes, and made to produce proteins on command. These operations are so efficient that yeast geneticists have gained a head start over most other groups.

THE YEAST GENOME

About a decade ago, several European labs decided to cooperate on an ambitious plan: They would decipher all the DNA in the entire yeast genome.

A Brief Key to Basic Genetics

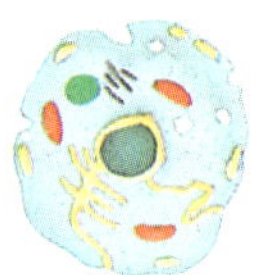

A human cell. *Each of the 100 trillion cells in the human body (except red blood cells) contains the entire human genome—all the genetic information necessary to build a human being. This information is encoded in six billion base pairs of DNA (three billion base pairs from each parent). Egg and sperm cells each contain three billion base pairs of DNA.*

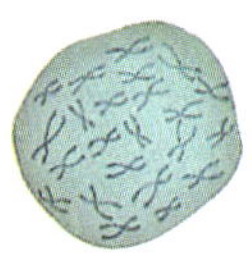

The cell nucleus. *Inside the cell nucleus, six feet of DNA are packaged into 23 pairs of chromosomes (one chromosome in each pair coming from each parent).*

A chromosome. *Each of the 46 human chromosomes contains the DNA for thousands of individual genes, the units of heredity.*

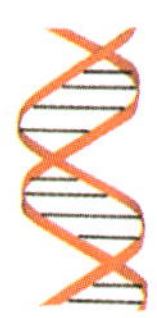

A gene. *Each gene is a segment of double-stranded DNA that holds the recipe for making a specific molecule, usually a protein. These recipes are spelled out in varying sequences of the four chemical bases in DNA: adenine (A), thymine (T), guanine (G), and cytosine (C). The bases form interlocking pairs that can fit together in only one way: A pairs with T; G pairs with C.*

A protein. *Proteins, which are made up of amino acids, are the body's workhorses—essential components of all organs and chemical activities. Their functions depend on their shapes, which are determined by the estimated 40,000 genes in the cell nucleus.*

After they divvied up the yeast cell's six chromosomes, each lab agreed to tackle a different chunk of them. The international effort eventually involved about 600 scientists, including some American and Japanese teams.

In 1996, the yeast community proudly published the result of its work: the complete set of genetic instructions for making a yeast cell—"a fantastic milestone," says Brown. These instructions were spelled out in 12 million base pairs, or subunits, of DNA encoding roughly 6,000 genes (genome-www.stanford.edu/Saccharomyces). "Whatever a yeast needs to do its job is somewhere in those 12 million or so base pairs," Brown says.

Until then, no one had ever sequenced the entire DNA of any eukaryote (an organism whose cells have a true nucleus and many other features of animal cells). This achievement marked the start of a grand voyage of discovery that is leading through the genomes of several other model organisms, including the roundworm (whose

Baker's yeast cells reproduce by budding and double every 90 minutes. Billions of them fit in a few petri dishes.

sequence was completed in December 1998) and the fruit fly (March 2000), right up to the gigantic, almost mythical human genome, whose sequence was announced—in rough draft—in June 2000.

Many of the genes in these simpler organisms bear an uncanny resemblance to those of humans. Unlike human beings, however, these organisms can be grown rapidly, and their genes can be mutated at will—a quality that researchers find essential, since mutations that stop a particular gene from functioning make it possible for scientists to figure out what the normal gene's function would be. Furthermore, model organisms can be mated in any way that suits an experimenter, and their offspring analyzed with ease. That is how scientists have deciphered several genetic pathways that are also found in human cells.

HOW SIMILAR ARE WE TO BAKER'S YEAST?

It is hard enough to accept that little creatures such as worms and flies are, as one researcher put it, "our relatives," but what about baker's yeast? What can we possibly have in common with this humble cell, really just a fungus, which is generally used to make dough rise or to brew beer?

The answer is stunning: We are so similar to yeast, in some of our genes, that human DNA can be substituted for the equivalent yeast gene—and it works just as well. This was first demonstrated in 1985, when Michael Wigler and his associates at Cold Spring Harbor Laboratory "rescued" a mutant yeast cell that lacked an essential developmental gene, the yeast equivalent of the human *RAS* gene, by inserting the human gene into it.

This "remarkable result," wrote David Botstein, a yeast researcher at Stanford, "indicated a profound conservation not only of [DNA] sequence but also of detailed biological function" over at least a billion years of evolution from yeast to human. More than 70 additional human genes have proved able to repair various mutations in yeast, he said. Botstein's conclusion: "What is true for yeast is also true for human."

Just like human cells, yeast cells go through cycles of growth and division—and the genes that regulate these cycles in yeast are practically identical to ours. In fact, nearly everything we know about the human cell cycle, and much that we know about human cancer, was originally learned in yeast (see p. 24).

Such knowledge used to be gained slowly and painfully. When scientists examined the newly sequenced yeast genome a few years ago, they were shocked to find that even though more was known about yeast than any other model organism, two-thirds of yeast's genes had never been identified before—and nobody had any idea what their functions might be.

"One of the great things about the genome sequencing projects is that they make it obvious we're still in the frontier days of biology," says Brown. "At the same time, these projects provide us with a very useful tool that allows us to explore large sets of genes and to discover their properties."

Most of the genes in organisms that scientists study "are really not understood at all in terms of their roles," he notes. "And for humans, it's 90 to 95 percent of the genes that we haven't a clue about. That's just a reminder of how much we don't know."

A MICROARRAY SHOWS HOW THE GENES IN A YEAST CELL RESPOND TO DIFFERENT TYPES OF STRESS

Brown's robot can be used to see the changing patterns of activity in a cell's genes when the cell is placed in different kinds of stressful circumstances. This microarray reveals that although thousands of genes in a yeast cell respond to each type of stress in a different way, about 900 of the predicted 6,200 genes in the yeast genome respond to all stresses in the same way.

During their common "environmental stress response," about 600 genes become less active while another 300 genes increase their activity. Each column represents the genes' responses to one particular type of stress. Each horizontal row represents a single gene's activity during varied stresses. The genes were grouped according to the similarity of their expression profiles.

Shades of red represent increases in mRNA expression, while shades of green represent decreases in mRNA expression. Black indicates there was no measurable change compared with unstressed cells.

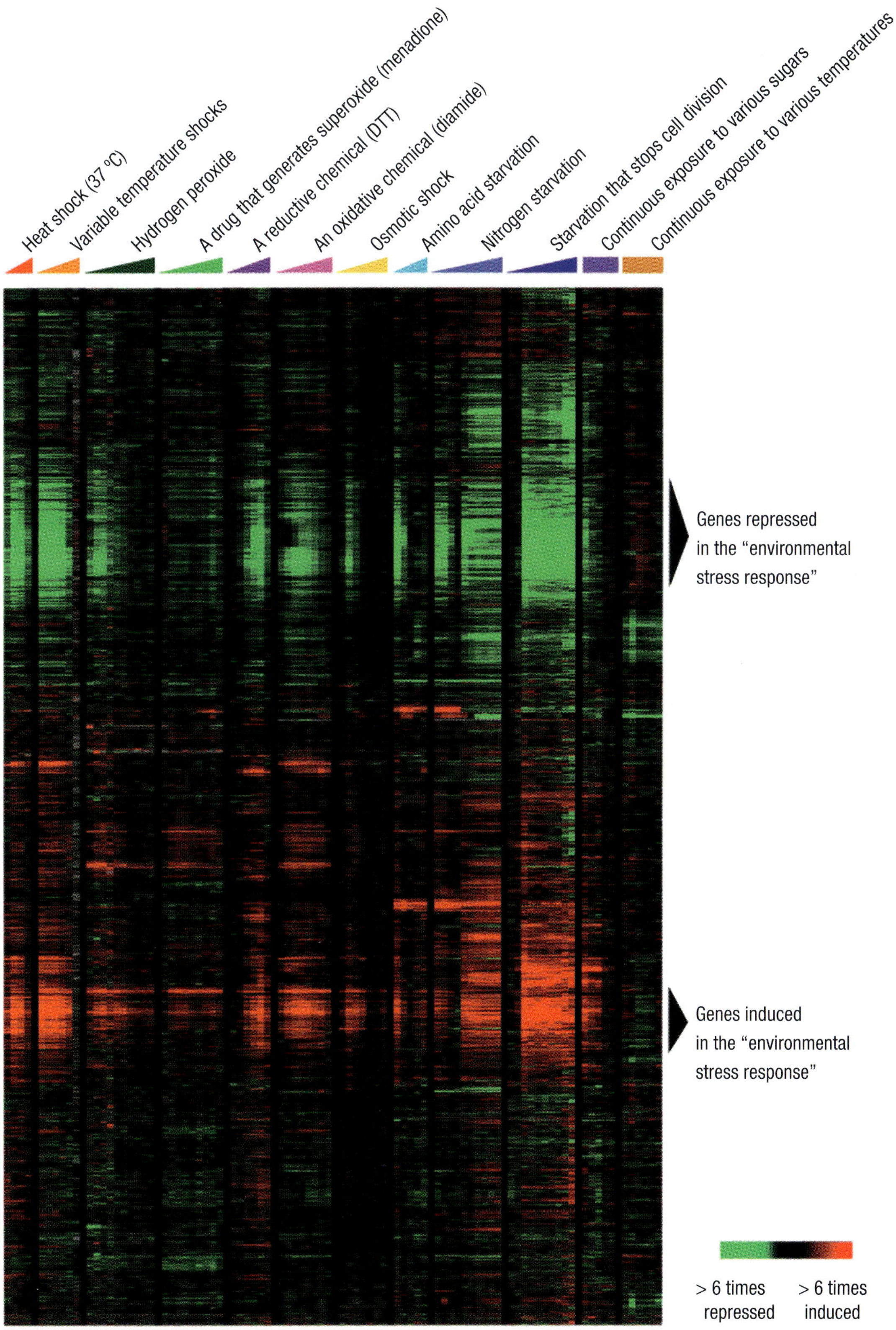

Laboratory of Patrick Brown, Stanford

HOW A MICROARRAYER WORKS

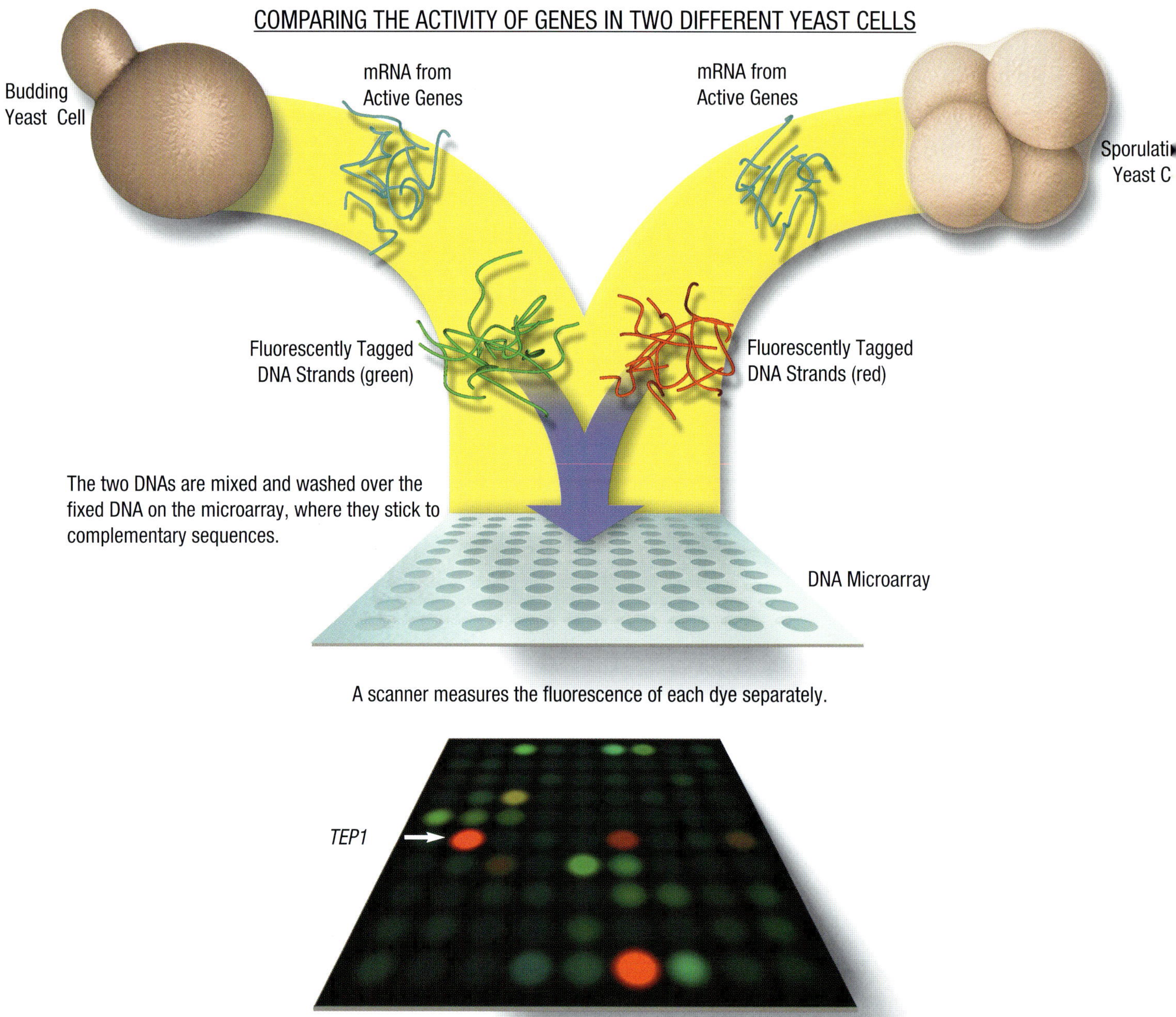

The higher ratio of red to green fluorescence at the spot labeled *TEP1,* for instance, shows that the *TEP1* gene is expressed more highly in sporulating than in budding yeast cells.

THE MICROARRAYER

Brown and his associates recently developed a machine called a DNA microarrayer, with which they hope to make a big dent in this ignorance. Naturally, since the yeast genome was the first to be completed, the new microarrayer was used first with yeast. But the technology is spreading to the genes of many other organisms and is even being used with fragments of the human genome.

A microarrayer is basically a robot whose pointed tip can deliver up to 20,000 bits of DNA, each representing a different gene, to specific spots on microscope slides. These "printed" DNA strands remain fixed on the slides. When samples of fluorescently tagged DNA made from a cell's active

HOW A DNA STRAND FINDS ITS MATCH

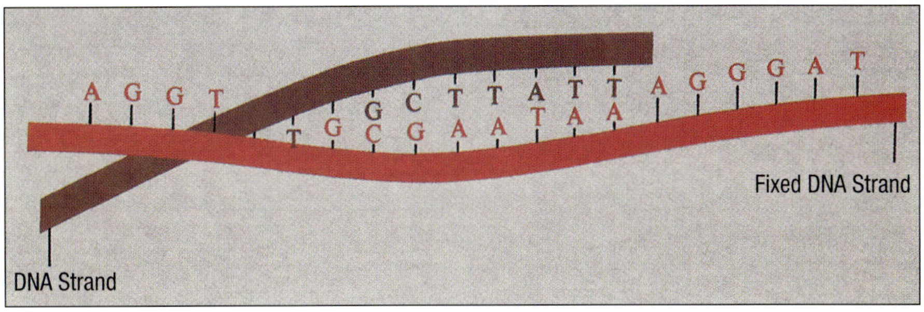

A strand of DNA from a particular gene (brown) sticks to a complementary sequence of DNA printed on the microarray (red). Chemical base A can pair only with T, and C can pair only with G.

genes are washed over the glass slides, any sample that complements a fixed strand of DNA will stick to it (the chemical base adenine [A] can pair only with thymine [T], while the chemical base guanine [G] pairs only with cytosine [C]), making it glow. The location of the glowing spots thus reveals which genes are active in the cell at that time.

The method works because the tagged DNA is a copy of messenger RNA (mRNA), which is produced only when a gene is "expressed" and actively making protein. The glowing spots therefore represent only the genes that are turned on in the cell at that particular moment. (While every cell contains all the genes in the genome, only some of the genes are turned on at any time, depending on the type of cell and its circumstances.)

The neatest part of this technique is that it enables researchers to compare the activity of genes in two different cells simply by tagging their DNA with two different types of fluorescent dye, which glow in different colors. The two sets of tagged DNA are mixed before being spread over the DNA that is printed on the glass slides. The spots where fragments of tagged DNA stick to the fixed DNA are illuminated, and scientists measure the fluorescence of each dye separately with the aid of an optical scanner; then they measure the ratio of one color to another. "The brighter green a dot glows, for instance, the higher the level of expression of a particular gene from the cell whose DNA is labeled green," Brown explains. "The brighter red, the lower that gene's level of expression compared to an identical gene from the other cell." When the two genes are expressed to the same extent, the dots glow in yellow.

Though "in some cases it may be okay to say that a gene is turned on or turned off," Brown adds, "more often it's a gradient. You need to measure the subtle differences in expression. So the color of the signal carries information about how vigorously the gene is expressed." As a result, scientists can see the activity of genes in different cells, or at different times, simply by examining the patterns of red, green, or yellow dots in the microarrays.

THE SEXUAL DEVELOPMENT OF YEAST

Brown used this technique in 1998 to study the sexual development of yeast in collaboration with Ira Herskowitz at the University of California, San Francisco. ("Actually, our graduate students, Shelley Chu and Joe DeRisi, did most of the work and thinking," explains Herskowitz.)

When one looks at highly magnified pictures of yeast, these little blobs seem utterly sexless. Yet yeast comes in two mating types—"a" and alpha—as well as a slightly larger diploid cell that carries both "a" and alpha. When a diploid cell runs out of food, it morphs into spores of the two different mating types, which remain enclosed in a tough shell until times improve. Then these spores burst forth and start looking for mates of the opposite type; as soon as they find a suitable mate, they fuse together.

YEAST SPORES AND HUMAN SPERM: A PARALLEL DEVELOPMENT

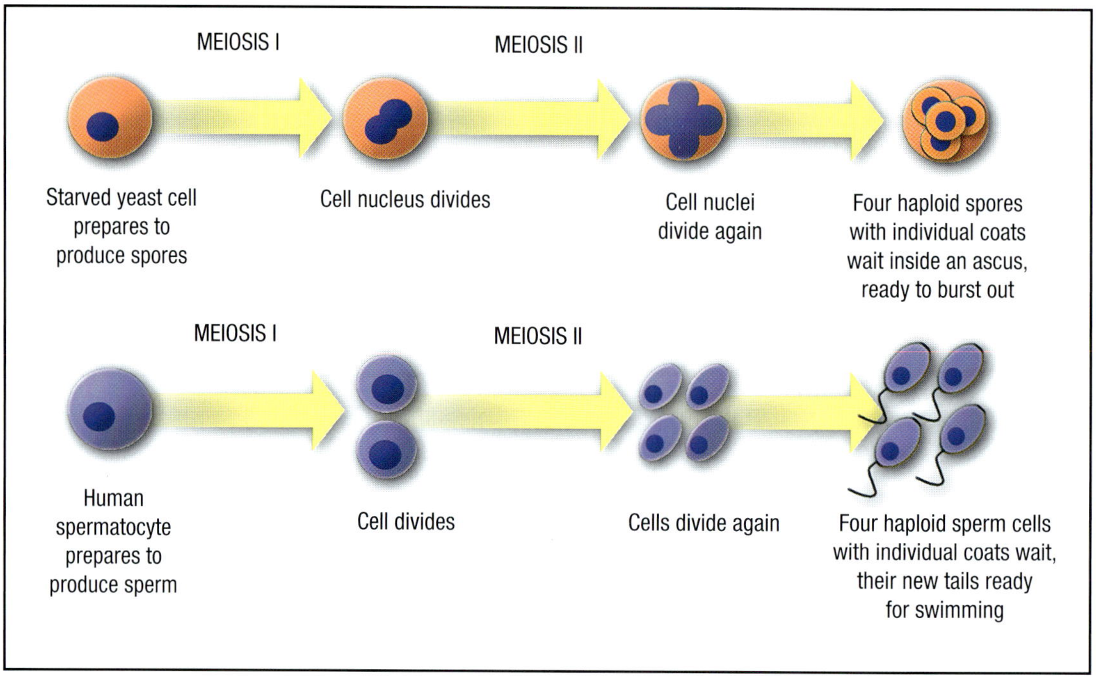

The spores' development is surprisingly similar to the development of human sperm and eggs. "Most of the cells in your body are diploid. They have two complete copies of the genome, one from your mother and one from your father," Brown explains. "But egg or sperm cells carry only one copy of the genome. To produce these sex cells, the diploid cells have to go through meiosis—they divide—and shuffle their chromosomes around.

"Well, yeast does exactly the same thing when its diploid cells form spores," Brown says. "Many of the genetic mechanisms are shared with humans. Many of the genes that are involved in the process, and the gene products, are functionally analogous to one another."

Until the Brown and Herskowitz study, scientists knew of roughly 150 yeast genes involved in making spores. But the microarrays, which show all the genes that are active in a cell at a particular time, gave the team a much richer, more detailed picture of what turns out to be a surprisingly complex process. Instead of 150 genes, they found that "more than 1,000 yeast genes showed significant changes in mRNA levels" during sporulation (the process of making spores). The activity of half of these genes was turned up (as shown in graded shades of red), while that of the other half was turned down (green) in complicated sequences that changed over time.

Around 10 percent of the changes in the genes' activity were in some sense expected, since scientists knew something about the genes involved. "But there was a large bunch of genes about which we knew essentially nothing," Brown says, "and here we suddenly had a really great clue as to what they might be doing."

GUILT BY ASSOCIATION

The clue? "Guilt by association," the team called it. "We can guess their function based on the fact that other genes—whose function we do know—are turned on at the same time," Herskowitz says. Genes tend to be expressed just when their products have important roles to play, he explains, so all the genes whose mRNA went up or down at the same time were likely to have similar functions.

These assumptions could be tested very easily in yeast. "We took a set of genes that showed a suggestive pattern of expression during sporulation," says Brown. "Then we knocked these genes out one by one and found that the cells failed to complete their program." The program stopped at various points, depending on which gene was knocked out. Each point revealed the normal function of the knocked-out gene.

"This study is a door opener; our analysis only scratches the surface," says Herskowitz. A searchable database of all the data from this project is now available on Brown's Web site (cmgm.stanford.edu/pbrown), where anyone can apply the information to human sperm and eggs (whose

How do yeast cells mate? Ira Herskowitz explains it all: An odorant substance called alpha-factor (arrow) released by an alpha cell binds to a receptor on the surface of an "a" cell (the opposite mating type). This sends a cascade of signals into the "a" cell, leading it to turn on several genes that are involved in mating, and eventually the two cells fuse.

The signal pathway is "just like those used in human brain cells to control the function of the brain," says Herskowitz.

formation, according to Herskowitz, is still "very poorly understood") or to analyze other developmental processes.

Over the past year, Brown and his colleagues alone have done at least 800 different measurements of how every single gene in yeast is expressed under varying conditions. They have moved in other directions, too. The tools that have been developed for analyzing genomes are "all-purpose," Brown points out; they can be used with any organism. Working with Gary Schoolnik and others at Stanford, Brown applied them to the bacterium that causes tuberculosis, *Mycobacterium tuberculosis,* whose DNA sequence had just been published.

"The minute a genome is sequenced, you can have an experimental tool. Bang! It's fast," says Brown. "We've already printed an array of this one. We've even done experiments looking at the response of *M. tb* to various antibiotics and trying to get an idea of the genetic basis for its high virulence."

In addition, Brown's coworkers have helped Stuart Kim and other Stanford scientists who study *C. elegans* make a microarrayer of the worm's genes (see p. 71). Human genes are not far behind. In fact, work with human genes started even before a rough draft of the gigantic human genome had been completed.

"As people look at large-scale pictures of the expression programs in genomes, they've begun to realize that there's at least as much information in genomes entirely devoted to controlling where and at what level the genes are expressed [as to defining protein end products]," Brown points out. Gene expression is what really distinguishes one cell from another, he says, "and suddenly, that's just an open book."

A MORE PRECISE VIEW OF BREAST CANCER

"One of the most awesome uses of the microarray is to characterize different kinds of cancer cells," Herskowitz declares. "Suppose you are treating a group of 10 women who all have breast cancer and their cancer cells look similar. If you could use a DNA chip with 5,000 known human genes and ask which genes are up and which are down in various patients' cells, you might find out that these 10 women actually have several different kinds of cancer.

"And this is incredibly important," Herskowitz continues, "because it allows you to give them very precise therapy. Assume their cancer cells fall into three different patterns of gene expression: A, B, and C. If you give all these women a particular regimen of drugs, you might find that women with type A breast cancer are very effectively cured by it, while for the others it's useless. So if you treated those women as one group, you'd say, 'Well, we had a 3 out of 10 success rate,' which people would say, 'Well, that's pretty good.'

"But it was actually better than that—3 out of 3 for the right disease, and zero out of 10 for the others. So this kind of information is going to revolutionize drug discovery," says Herskowitz, "because now you can do clinical trials in which you treat people who have exactly the same disease." At present, most of the scientists in Brown's lab do experiments both in yeast and in human cells. "It's more fun that way," says Brown.

Brown's microarrays are not the only kind available to researchers. There are several different types to choose from, including the popular DNA chips made by Affymetrix, a biotech company in Santa Clara, California. These chips have much in common with silicon chips, and the technique with which they are produced—photolithography—is normally used in the semiconductor industry. Each chip is a half-inch square of glass on which thousands of short filaments of DNA have been imprinted. The DNA filaments are synthesized from lab chemicals and represent known sequences of DNA. When a liquid that contains chopped-up genes (or mRNA) from a particular cell is poured over the chip, only those bits of genetic material that perfectly complement a synthetic filament on the chip will stick to the chip and glow. A scanner then reads out their pattern.

Factory-made DNA chips such as this one are produced with techniques normally used in the semiconductor industry.

SPREADING THE BENEFITS

The DNA chip technology has certain advantages, but it is still quite expensive, requiring an outlay of at least $200,000 plus the cost of additional chips. To bring whole-genome analysis within the reach of all researchers, Pat Brown recently posted a do-it-yourself guide on his Web site. The guide provides a complete list of necessary materials for his type of microarrayer. It also shows where to order these parts and explains how to put them together.

The basic equipment for the microarrayer plus the scanner costs less than $80,000, Brown says. Printing a microarray that contains the entire yeast genome now costs as little as $10, since the DNA is spotted onto conventional microscope slides. Hundreds of copies of such microarrays can be printed in just one day.

Thousands of scientists have logged on to this site in recent months. More than 100 homemade microarrayers are now in operation in labs around the country, and Brown has started to run hands-on courses through which many more researchers may learn to build their own.

Being able to do experiments with microarrays is "the main immediate benefit of having genome sequences," Brown says. He wants to make sure that this benefit spreads far and wide.

DISCOVERING WHAT THE GENES IN A LIVING CELL DO—STARTING WITH YEAST

IT'S COOL TO MAKE A

They were a happy but slightly nervous group of scientists—16 handpicked lab chiefs, one of whom had come all the way from New Zealand to Long Island, New York, for the experience. They would build DNA microarrayers from off-the-shelf parts under the eyes of the machine's inventors, Pat Brown and his associate Joseph DeRisi, and learn how to use the machines efficiently.

They met in a new building at Cold Spring Harbor Laboratory, which hosted the two-week program. "People shouldn't have to wait for years and years for this new technology," declared James Watson, codiscoverer of DNA's structure and president of the laboratory, as he wandered through the messy room early in the session. "If you can look at 15,000 genes at one time, instead of 3, you can make so much more progress.... After I heard Pat Brown speak at a meeting recently, I thought, 'You have to have a course!'"

Eight times more scientists had applied for this training than could be admitted. Now the chosen few—12 men and 4 women—were faced with mounds of crates and boxes full of bewildering parts that they were expected to assemble into four working machines. Clearly uneasy, they formed small groups around piles of equipment.

Within hours of arriving in the lab, the students have assembled the robot's two horizontal stages and start wiring up its control circuitry with help from Joe DeRisi (third from left) and the printed "MGuide."

New Zealander Don Love of the University of Auckland assembles the robot's vertical stage, an important part that will hold the printhead. Sudha Veeraraghavan of the University of Texas, Houston, looks on.

Considering their next move, Tom Kocarek of Wayne State University, Detroit, and Michelle Marron of the University of Florida, Gainesville, seem deep in thought.

MICROARRAYER

Many of them kept consulting their copies of the "MGuide," a loose-leaf book that had been printed off Pat Brown's Web site. Brown had posted the step-by-step instructions for making DNA microarrayers on the Web so that his techniques would be freely available to all researchers. Nevertheless, the visiting scientists seemed glad to have Brown and DeRisi right beside them as they started their work.

DeRisi, in particular, knew most of the pitfalls. He had actually built a prototype of this arrayer himself a few years earlier, during three months of intensive work. "I got the parts from various manufacturers, but not a lot of literature gets shipped with these parts," he recalled. "Eventually I figured it all out and kept a *big* notebook." Since then, he has made five more arrayers, improving and simplifying the procedure on each pass.

The parts for an arrayer cost around $25,000. Brown and DeRisi had sent four sets to Cold Spring Harbor Laboratory from Stanford. They also

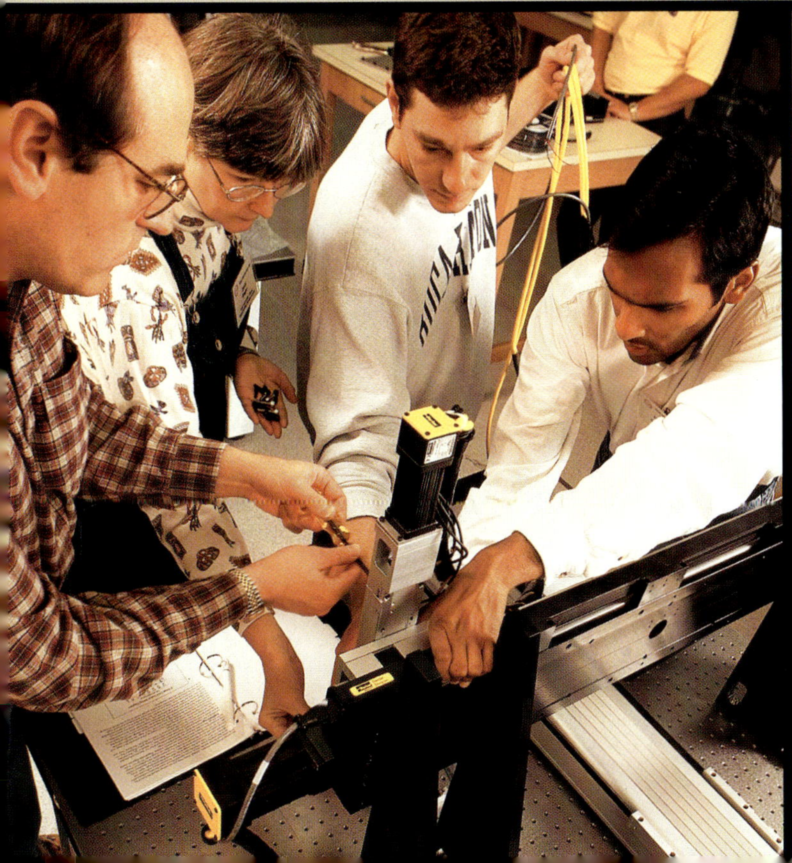

◀ *Four scientists from Princeton University, Yale University, the New York State Department of Health, and the University of Michigan work as a team to attach the vertical stage to the microarrayer.*

▲ *Alarmed by a loud grinding noise, Joe DeRisi shouts instructions about the emergency stop button and adds, "Remember that!"*

brought a container of yeast genomic DNA so that each visiting scientist could amplify (make multiple copies of) it and learn how to print it onto glass microscope slides in the new arrayers.

The early morning chaos of wires, cables, motors, odd-shaped parts, and wrappings on tables and on the floor slowly cleared up. By 2 p.m. on the first day, most of the arrayers' foundations—three servomotor-powered linear rail tables over an antivibration table—had been laid and the participants seemed a little less lost. Brown was impressed. "Things are running amazingly smoothly," he remarked. "There have been no glitches! Usually there are some. I kinda wish we had some, because it's a great troubleshooting experience to find out what went wrong."

His wish was soon granted. Suddenly a loud, ominous grinding noise brought everything to a standstill. "Ooh, what did I do?" exclaimed a male voice. Another scientist immediately pushed a red button in a square yellow box on the table and the sound stopped. DeRisi jumped into the fray. "If you need to hit the emergency-stop button, you must power down the amplifier and turn down the program before you release this button," he warned loudly. "Remember that!" He quickly identified the problem and took care of it, reassuring the perpetrator that there had been no serious damage.

The scientists picked up where they had left off. At one table, four of them spent more than three hours doggedly aligning and realigning the printhead until it fit correctly—an essential step. The whole group seemed increasingly determined to keep going. After dinner they returned to the lab for more work. By 12:30 a.m. that night, three of the arrayers were up and running (the last one turned

Sudha Veeraraghavan carefully places a glass slide on the arrayer platform in preparation for her group's first print run.

out to have a few missing parts, which would arrive shortly.)

UP AND RUNNING

"I'm not an engineer. I'm a geneticist," commented Donald Love of the University of Auckland, New Zealand, over breakfast the next morning. "I thought it would take at least a week to assemble the microarrayer. But we got it set up the very first day!"

The second day was even better, as the scientists began to install the microscope slides and to assemble the printing robots, each with 16 vertical, sharp-tipped pins that would deliver DNA "spots" to precise locations on the slides. After lightly touching each slide, the 16 pins would be dunked into a cleaning solution and dried before picking up fresh loads of DNA that represented other genes. Then the robot would maneuver the pins to a slightly different position over the glass slides and, acting like the tips of fountain pens, the pins would deposit extremely small, measured volumes of the new DNA onto new spots. "This is done over and over again, with very slight differences in the position of all 16 pins. That's how you create a high-density array," explained one of the participants as if he were an old hand at it.

Everything was ready at one of the lab tables. A scientist flipped a switch, and suddenly the pins started going up and down together, making a regular "choo-choo-choo" sound like a toy train engine. "It's happened!" cried out

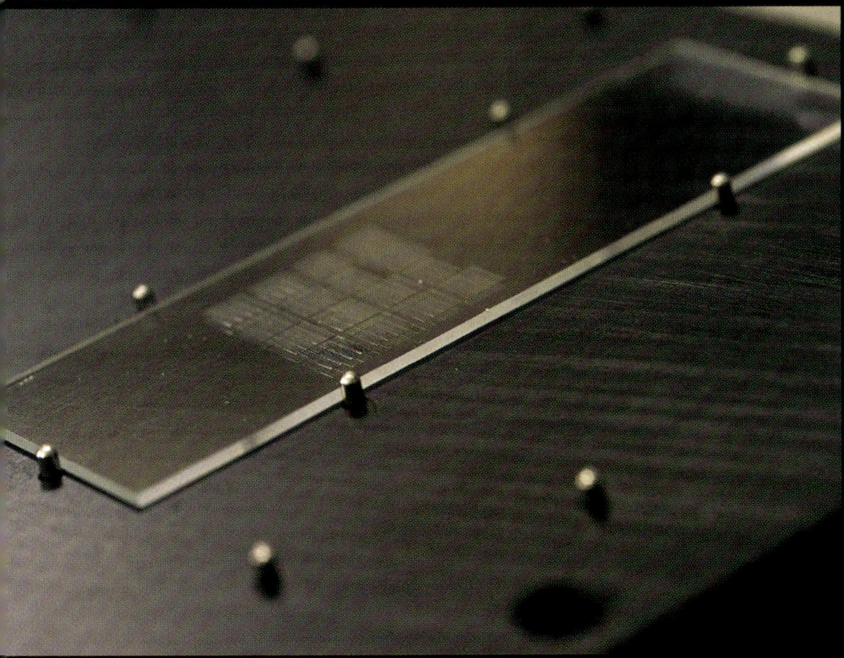

The printhead—an array of 16 sharp-tipped steel pins that act like fountain pens—makes contact with a glass slide and prints 16 tiny spots of DNA solution, each one representing a different yeast gene.

▲ *6,400 different genes are printed on this slide. A pattern of 16 squares, each composed of 400 tiny spots, is visible because of salt residue from the evaporation of the printed DNA solutions.*

Pat Brown and Arul Chinnaiyan of the University of Michigan check out a slide from a test print run. So far, so good....

one of the group excitedly. "That's great," said DeRisi. "Once you get to the point where you're aligned and printing, you're in steady state. It just keeps running."

By the third day, the participants were amplifying DNA of the entire yeast genome. Then they took turns actually printing it onto the glass slides and examining the results. "This is cool," said Michael Ryan of the New York State Department of Health as he held up a slide to the light. "It's the first one that printed!"

The rest of the course focused on how to use the new machines. The students designed and carried out several experiments. They washed samples of DNA over the printed spots on the glass slides to see where the DNA would bind; binding indicated that a particular gene was expressed in the cell that provided the samples. They saw that various genes changed their levels of expression in different circumstances. And they learned how to measure and interpret these differences, using a special scanner for measurement and new software developed by Michael Eisen, a bioinformatics expert at Stanford, to identify patterns in the data. Eisen, who had already made his software available to all scientists, had come to Cold Spring Harbor Laboratory specifically to teach the participants how to use it.

USING THE ROBOTS

The participants had many different reasons for wanting to master these techniques. "At least 20 genes are involved in juvenile diabetes, but for 18 of them we know almost nothing—we know only their general regions on chromosomes," said Michele Marron of the University of Florida immunology department. "Most of these DNA

Sudha Veeraraghavan, Maria Lagerstrom-Fermer of Sweden, and Vivek Mittal of Cold Spring Harbor Laboratory rejoice at the success of an alignment test used to fine-tune the printer's software settings.

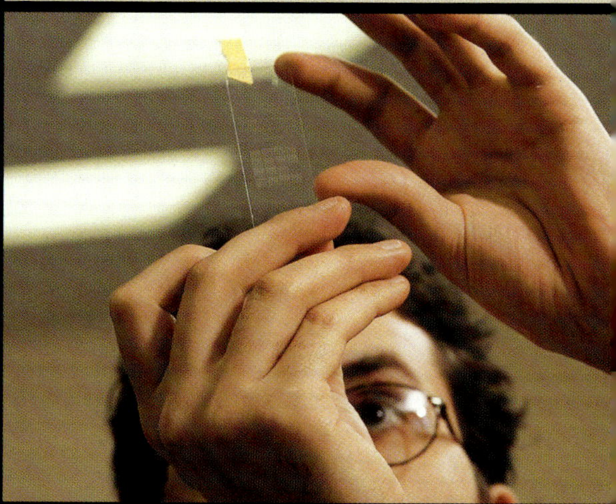

Inspecting a freshly printed slide, Ash Alizadeh, an instructor, holds it up to the light.

regions are very long and don't contribute a great deal. We hope to identify the genes by comparing levels of gene expression in normal and diabetic mice. We also want to identify the genes' pathways."

Christian Kaltschmidt of the University of Frieburg, Germany, who works on the mechanisms of memory, was eager to hunt for genes that may protect against Alzheimer disease. Matthis Lorenz of the National Cancer Institute in Bethesda, Maryland, is studying breast cancer. Maria Lagerstrom-Fermer of the University of Uppsala, Sweden, will compare gene activity in the cells of patients who have chronic myeloid leukemia during and after treatment.

Bruce Futcher of Cold Spring Harbor Laboratory will use the microarrayer he just built in the course to expand his work on the cell division cycle. In 1998, he collaborated with Paul Spellman, Pat Brown, and David Botstein to locate 800 yeast genes that are regulated by this cycle. Now he wants to study the clusters of genes that are turned on, one after another, during this cycle and to identify the functions of individual genes in the clusters.

In addition to doing their own research, nearly half the program participants intend to set up core facilities that will make and provide arrays for other scientists in their universities, hospitals, or research centers.

Once a microarrayer is set up, "it's so easy and cheap to make these things," says Pat Brown with missionary zeal."You could print 100,000 arrays per year with four machines! It's so routine.... I wanted them to see it. I wanted to make sure they would not find it intimidating."

Arul Chinnaiyan and Max Diehn, an instructor, examine a series of test prints.

James Watson (far right) discusses the students' progress with Pat Brown (center) and Bruce Futcher.

DISCOVERING WHAT THE GENES IN A LIVING CELL DO— STARTING WITH YEAST

AN AWESOME POWER

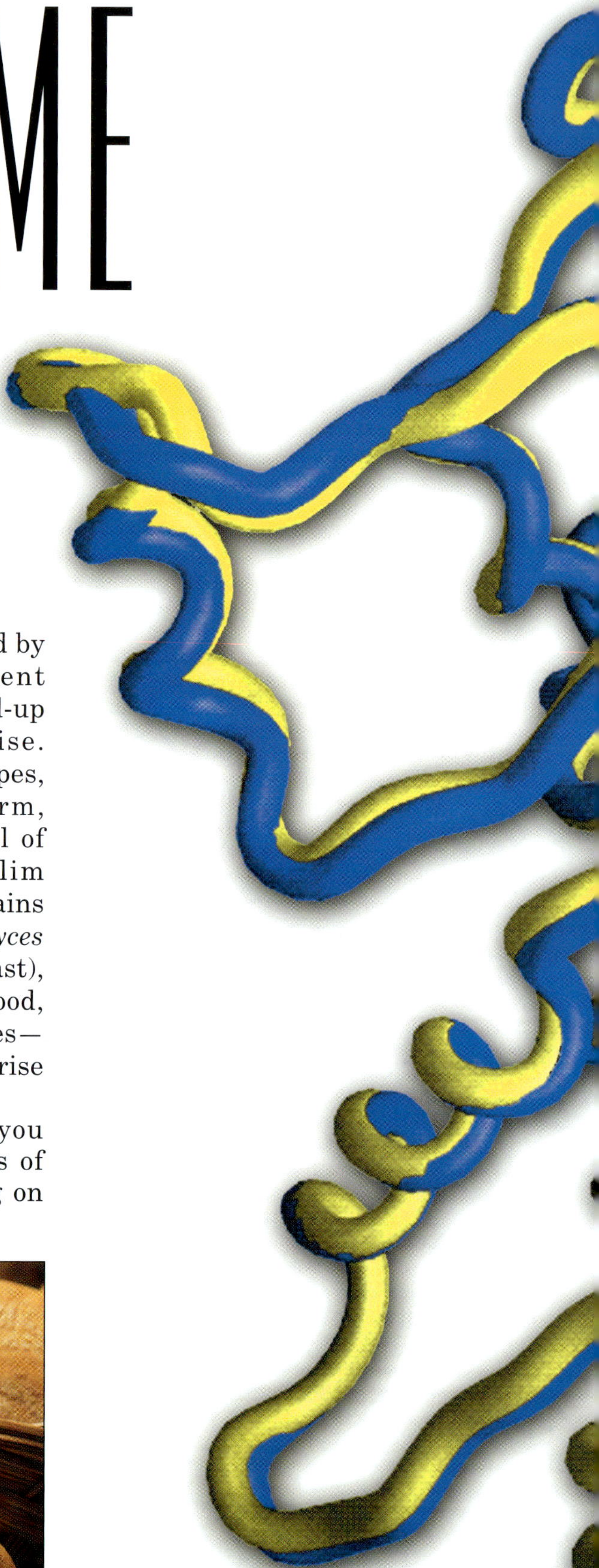

Yeast has been known and loved by human beings since ancient times. When added to mashed-up wheat, it makes bread rise. When added to squashed grapes, it produces wine. Its warm, slightly fruity smell is the smell of live cells at work. Even a slim package of "active dry" yeast contains roughly 35 billion cells of *Saccharomyces cerevisiae* (baker's or brewer's yeast), and as soon as these cells find food, they produce fermentation bubbles—the bubbles that cause dough to rise or beer to foam.

Yeast is also plentiful. When you eat grapes, you swallow millions of live yeast cells that were feasting on

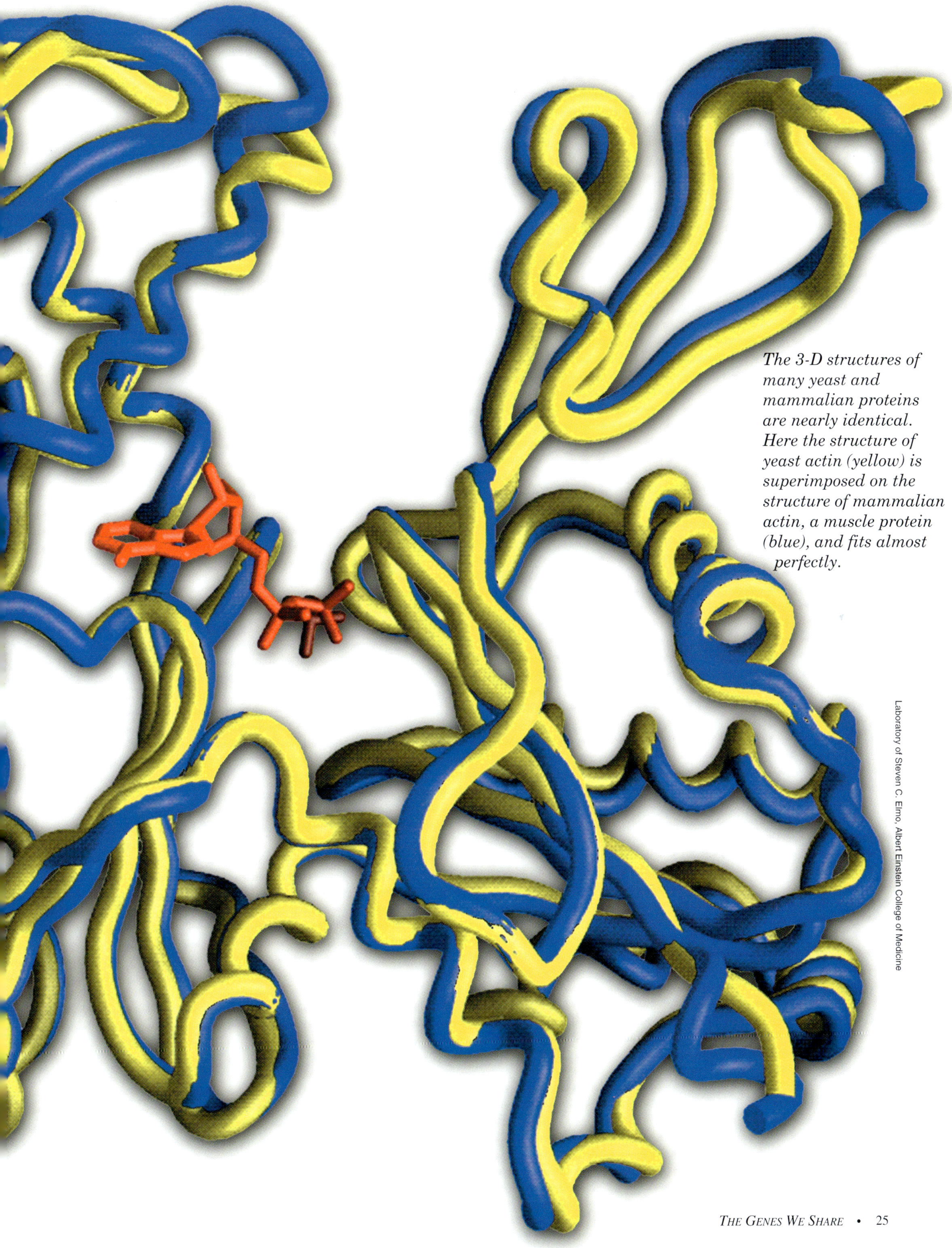

The 3-D structures of many yeast and mammalian proteins are nearly identical. Here the structure of yeast actin (yellow) is superimposed on the structure of mammalian actin, a muscle protein (blue), and fits almost perfectly.

Laboratory of Steven C. Elmo, Albert Einstein College of Medicine

sugar on the grapes' surface. Yeast is cheap. It's easy to grow—yeast cells double every 90 minutes when food is available, and billions of them will fit in a few petri dishes.

Not only are these cells alive (until oven heat kills them), they can do almost everything that human cells do to survive: transmit signals from the cell's surface to its nucleus, manufacture thousands of proteins, create a cellular scaffolding, repair DNA in the nucleus, and so on. And although yeast cannot make tissues and does not have a brain, it is universally recognized as the best model organism for studies of anything that goes on inside a single cell.

Researchers speak glowingly of the "awesome power" of yeast genetics to solve problems in biology. "It's so easy with yeast!" says Randy Schekman, an HHMI investigator at the University of California, Berkeley. He points out that "genetic surgery has been possible in yeast for 20 years. You can stitch bits of

FINDING A GENE IN THE YEAST GENOME

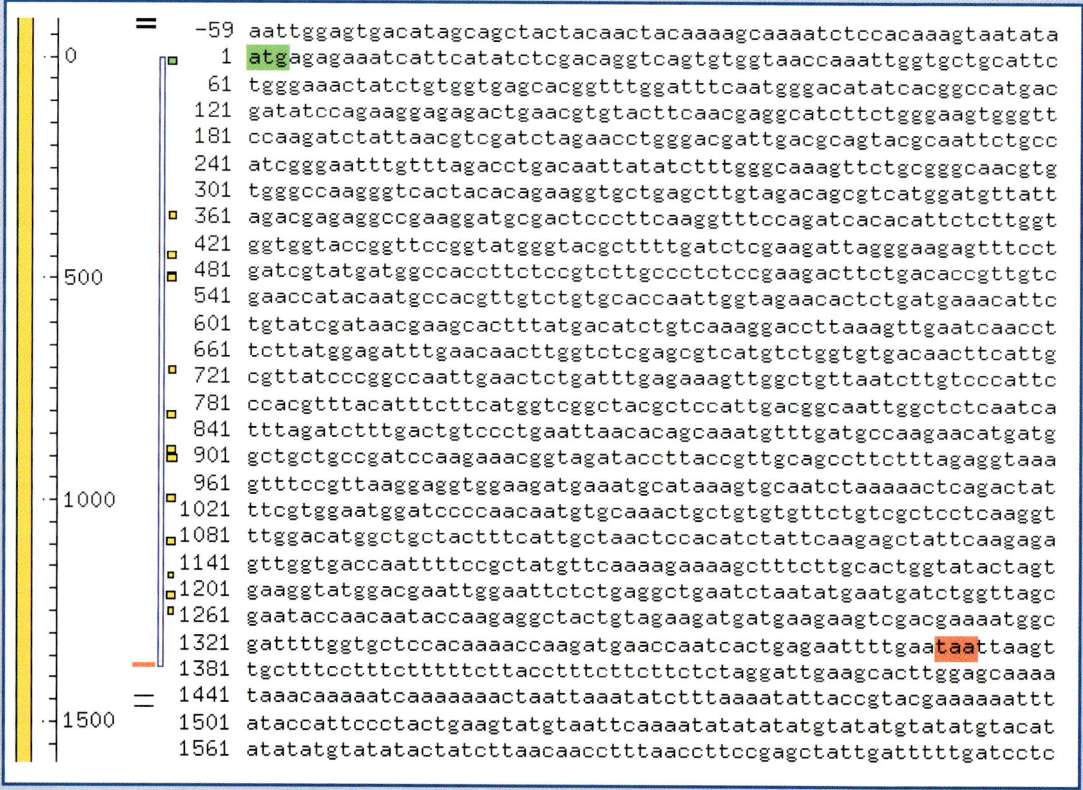

"It's easy to recognize a gene in the yeast genome, compared with the human genome," says Pat Brown, "because the coding sequences of yeast genes are generally uninterrupted."

You look for a start signal—the sequence ATG. This must be followed by an open reading frame (ORF), a long stretch of consecutive triplets of DNA bases in which each 3-base chunk of DNA codes for a single amino acid. These triplets (codons) must not include any stop signal (either TAA, TAG, or TGA).

Searching for potential genes in a 1,500-base segment of yeast chromosome 6 (right), Stanford scientists picked out a start signal (green) and then read the remaining DNA in triplets until they reached a stop signal (red). This identified the gene for beta-tubulin.

SIGNALS MUST BE "IN FRAME"

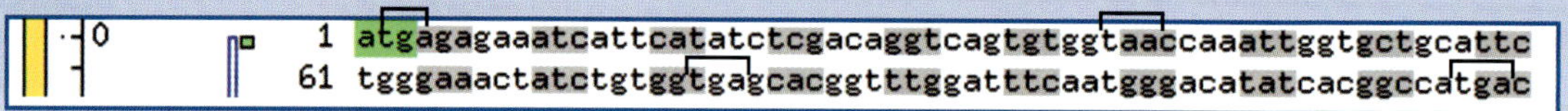

Scattered throughout the genome are hundreds of combinations of DNA bases that could be stop signals if they were "in frame" but are ignored because they do not form consecutive triplets after the start signal. Some examples of these out-of-frame triplets are indicated by brackets above.

DNA from a normal yeast cell onto a plasmid, the plasmid will carry this DNA into a mutant yeast cell, and the normal yeast gene will replace the mutant one at precisely the right place."

It's also very quick. The yeast genome has so few genes—6,000, compared with the estimated 40,000 genes in mice and humans—and the tools for manipulating these genes are so highly developed that yeast experiments can generally be performed in days or weeks. Similar experiments in mice might take years—if they could be done at all.

Another great virtue of yeast is that its genes have very few of the bothersome introns (intervening sequences of DNA) that interrupt the coding sequences of mammalian genes. This makes it "easy to recognize where the genes are in yeast and what their boundaries are, just by looking at the genome sequence," Pat Brown explains. He notes that approximately 70 percent of the yeast genome codes for protein—a huge amount, compared with only 5 percent of the human genome. (The function of the remaining 95 percent of the human genome is still unknown.)

Despite these differences, large numbers of yeast genes closely resemble those of mice and humans. They also produce nearly identical proteins. "You can generally take a yeast protein and an equivalent human protein and superimpose their 3-D structures, and they will be crystallographically the same," says David Botstein. In his studies of actin, a protein, Botstein found that "not only is actin conserved [unchanged by evolution] from yeast to humans but most of the proteins that interact with actin in yeast are also conserved, and so are their interactions."

Philip Hieter of the University of British Columbia in Vancouver, a pioneer of what he calls "comparative genomics," remembers that when he gave talks about yeast in the 1980s, "there was a lot of skepticism. People didn't appreciate how similar yeast was to mammals. But the next 10 years showed that all the basic pathways within cells are much more conserved than anybody expected at the time. So if you examine a new human protein—a protein involved in a human disease, let's say—and you find that it matches a yeast protein, this brings up the entire biochemical pathway." For a biomedical scientist, that's like hitting a jackpot. ●

THE GENETIC CODE

Second base of codon

First base of codon	U	C	A	G	Third base of codon
U	UUU } Phe UUC UUA } Leu UUG	UCU } Ser UCC UCA UCG	UAU } Tyr UAC UAA UAG	UGU } Cys UGC UGA UGG Trp	U C A G
C	CUU } Leu CUC CUA CUG	CCU } Pro CCC CCA CCG	CAU } His CAC CAA } Gln CAG	CGU } Arg CGC CGA CGG	U C A G
A	AUU } Leu AUC AUA AUG Met	ACU } Thy ACC ACA ACG	AAU } Asn AAC AAA } Lys AAG	CUU } Ser CUC CUA } Arg CUG	U C A G
G	GUU } Val GUC GUA GUG	GCU } Ala GCC GCA GCG	GAU } Asp GAC GAA } Glu GAG	GGU } Gly GGC GGA GGG	U C A G

This ingenious dictionary, developed in 1966, shows how to recognize the 20 amino acid molecules that are the building blocks of proteins. Each amino acid is spelled by a codon—a triplet of letters, or bases, of either DNA or RNA. However, some amino acids can be spelled in more than one way, by several different codons.

The bases labeled U (uracil) in the genetic code above are RNA versions of the base called T (thiamine) in DNA. Codons may start with either U (T), C, A, or G. At the top left, for instance, the triplets UUU (TTT) and UUC (TTC) both code for the amino acid phenylalanine. At the bottom right, all four codons specify the amino acid glycine. The start signal AUG (ATG) also codes for the amino acid methionine.

CANCER: CLUES FROM YEAST'S

Yeast can afford to be selfish. As a single-celled organism, it need not worry about the welfare of any other cell. As long as it has enough food—ripening grapes, sticky leaves, or almost anything sweet—it just eats and reproduces every 90 minutes.

DISCOVERING WHAT THE GENES IN A LIVING CELL DO—STARTING WITH YEAST

Only abnormal yeast cells stop growing while their food supply is still plentiful. In the 1970s, Leland Hartwell of the University of Washington, Seattle, seized on this fact for his groundbreaking studies of the basic cycle through which a cell grows and divides into two daughter cells.

Hartwell chose to work with baker's yeast because it grows by budding. Unlike the cells of fission yeast, which reproduce by becoming larger and larger and then dividing in half, the cells of baker's yeast never grow bigger than they are at maturity. Instead, a little knob, or bud, starts to protrude from the cell's surface, breaking its neat symmetry, as the cell enters its reproductive cycle. A mere pimple on the cell's surface at first, the bud soon grows and balloons out until it is almost as large as the mother cell. Then it splits off.

Any mutation that affects cell division will make the bud stop growing after it has reached a particular size—an easy symptom to recognize. The bigger the bud when growth stops, the farther the cell had progressed along the division cycle when the effect of the mutation kicked in. Cells whose buds stop at different sizes therefore contain mutations that affect different stages in the cell cycle.

Using bud size as their guide, Hartwell and his colleagues succeeded in identifying the first mutant gene (*cdc*, or cell division cycle mutant) that was clearly responsible for blocking the cell cycle. Then they identified another 32 genes that were involved in halting the growth of yeast buds at different stages.

What made these findings riveting to researchers was that human equivalents of the yeast genes were soon found. Even more interesting, these human genes turned out to play crucial roles in the development of cancer—though cancer does not occur in single-celled organisms such as yeast.

Cancer results from cell division gone wild. Billions of cells in the adult human body divide every day to replace cells that die. Most of these cell divisions go perfectly well, thanks to an intricate system of checks and balances. But occasionally a cell will defy normal restraints on growth and begin to multiply uncontrollably, giving the body a strong push on the way to cancer.

"A single mutation in a cell's DNA is not enough to cause cancer," points out Bert Vogelstein, an HHMI investigator at The Johns Hopkins Oncology Center in Baltimore, who has tracked the development of colon cancer. "Cancer arises from the accumulation of several mutations," he says. Some

CELL-DIVISION CYCLE

A yeast cell grows a new bud (right) despite a scar from a previous birth (left). After dividing 20–30 times, yeast cells are covered with scars and cannot divide again.

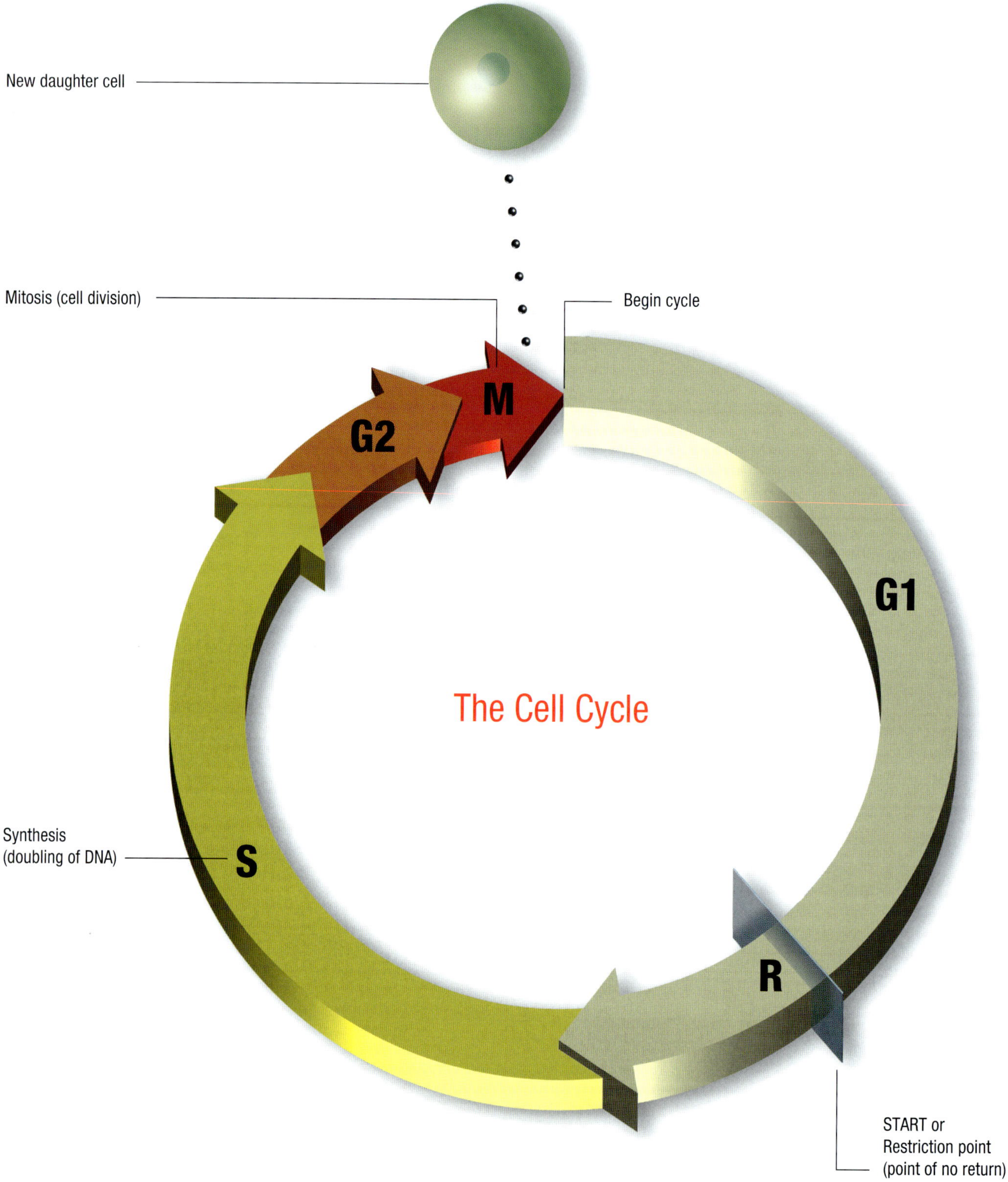

Before dividing, a cell must pass several checkpoints. First it enters a phase called G1, during which various signals determine whether the cell is allowed past a key point—START in yeast cells, or R in mammalian cells. Beyond this point, a cell is committed to entering the S phase and duplicating its entire DNA. Then the cell goes into the G2 phase and proceeds to mitosis, M, a very brief but dramatic process that results in the formation of two identical cells.

genetic mutations are inherited and produce predispositions to specific types of malignancy. Other mutations are caused by environmental factors such as radiation or poisons. These alterations in a cell's DNA may turn normal genes into dangerous oncogenes that accelerate cell growth, or they may inactivate "tumor suppressor" genes that would normally stop unwanted growth. Either way, a cell that has undergone enough mutations may start to proliferate so rapidly that its descendants gain a selective advantage over normal cells, which divide less frequently. Eventually the mutant cells form a tumor.

Scientists' best view of the normal cycle of cell division has come from research on yeast. The most dramatic stage of the cycle is mitosis (M), during which a cell's chromosomes, which contain its genes, are duplicated and two identical cells take shape. Mitosis was the first stage of cell division ever examined by biologists, although it is actually one of the shortest. Three other stages, all of which occur before mitosis, were identified in the 1950s: S, during which DNA is synthesized; G1, which appears even earlier in the cycle; and G2, the short period after DNA synthesis.

THE POINT OF NO RETURN

As Leland Hartwell suggested in the 1970s, there is also a START point, where the cell "decides" whether to replicate its DNA. Beyond this point, the cell is committed to entering the S phase. (In mammalian cells, this point is now called R, for restriction point.)

Passing through the restriction point is "like Caesar crossing the Rubicon—it's the point of no return," says Steven Dowdy, an HHMI investigator at Washington University in St. Louis. "When regulation is lost at this critical checkpoint in the cell cycle, you get malignancy or progress toward malignancy."

The cell division cycle has been compared to a circular highway that has signposts, checkpoints, exits, rest stops, and even detours. Cells that pass the restriction point and duplicate their DNA must face yet another checkpoint at the end of the G2 stage, when the duplicated DNA is examined for accuracy before the cell is allowed to enter mitosis. Both checkpoints are regulated by cdc2, the protein encoded by one of the genes that Hartwell identified in yeast.

After all the genes in yeast were sequenced in 1996, scientists took a fresh look at the cell division cycle and the genes that play a role in it. One team of researchers, led by Raymond Cho and Ronald Davis of Stanford University and their colleagues at Affymetrix, used DNA chips to monitor the activity of all 6,000 genes in a yeast cell as it progressed through cell division. Another team, led by Paul Spellman, Patrick Brown, David Botstein, and Bruce Futcher, used Brown's microarrays. Both teams identified several clusters of genes that turn on or off together during specific stages of cell division. The scientists are now searching for the regulatory elements that direct and coordinate this activity—elements that probably exist in humans as well.

Meanwhile, both teams have placed their data on the Web, in the expectation that genomic data sets of this sort will become increasingly valuable as more data accumulate. In such data, perhaps, lies the key to stopping cell division when it runs amok—and thereby preventing cancer.

THE "FLY PEOPLE" MAKE HISTORY WITH THE FRUITFUL FRUIT FLY

UNEXPECTED REWARDS FROM OUR SIMILARITY TO FRUIT FLIES

By
Gary A. Taubes

The first mutant fruit fly on record is a white-eyed fly that the legendary Thomas Hunt Morgan spotted in his "fly room" at Columbia University in 1910. But as early as 1907, Frank Lutz of the Carnegie Institution was remarking to his friends that his flies had "started to do tricks with their [wing] veins" (the prominent veins that form patterns on their wings), and in 1908 he observed dwarf flies among his cultures.

Since then, fruit fly mutants have proliferated like, well, flies, spawning an entire field of biology in the process. There are mutants with bizarre eye colors—pink, purple, maroon, or even vermilion instead of the normal brick red—and those with truncated wings, or miniature wings, or no wings at all. There are mutants that are very hairy and mutants that are bald.

There are giant embryo mutants and mutants with legs growing where their antennae or mouth should be. There are mutants with perfectly formed, functional eyes growing out of their wings, or on their legs, or at the tips of their antennae. There are mutants with two sets of wings instead of one, and mutants with no heads but two abdomens stuck end to end.

There are uncoordinated mutants and mutants with no memory. There are mutants that are hopeless at courting and others that court only their own sex. There are mutants that can't hold their alcohol. And these are only a tiny sample of mutant fruit flies.

By the end of the 1980s, fruit fly researchers—known as "fly people" or "Drosophilists," after the insect's Latin name *Drosophila melanogaster*—had catalogued some 3,000 different mutations, representing an extraordinary wealth of biological information. This value derived from a simple fact of genetic research: "The way you find out what a gene does," says Hermann Steller, an HHMI investigator and neurobiologist at The Rockefeller University, "is by generating a mutation and looking at the consequences—

Gerald Rubin, a leader in the effort to decode the fruit fly genome, stands before a small section of the completed sequence.

looking at what the fly does when that gene loses its function. If you had no idea what a car engine was doing or how it worked, you'd take out different parts to see what happens. This is the basic logic geneticists use to see what function genes have."

Over the course of a century, the fly people claim, this wealth of mutants has taught biologists more about the fundamental biological, developmental, and genetic processes of flies than about those of any other single complex organism. (Some researchers would argue that they know at least as much, if not more, about the roundworm *C.elegans*.) Drosophilists can now describe with a reasonable degree of precision how a fruit fly embryo is put together: what genes turn on when in the course of its development, what those genes do, and why they do it.

All this knowledge would have been of interest only to a small group of specialists except for one thing. The fruit fly work helped spark a revolutionary understanding in biology: that the fundamental genetic mechanisms of development appear to be the same in all living things. Once evolution stumbles upon a mechanism that works, says Matthew Scott, an HHMI investigator at Stanford University, it uses it over and over again. As a result, research on the genome of *D. melanogaster* is playing a pivotal role in uncovering biological processes that all organisms share.

"If you just have the sequence of the human genome," says Gerald Rubin, vice president of HHMI and professor of genetics at the University of California, Berkeley, who has been working on *D. melanogaster* for 25 years, "you won't know how to interpret it. It's written in a foreign language that you can't read. What you need to interpret it are the genomes of model organisms, in which people have already done a lot of work to determine what the functions of the genes are....

"You can't do the kind of experiments in humans that you can do in the model organisms. You can't say, 'I want to cross that person with that person and see what their grandchildren are like.' So you have to go back and look in these other genomes for clues to how the human genome works."

Researchers have been cloning the genes of *D. melanogaster*, one gene at a time, since the mid-1970s, when recombinant DNA technology first became available. But the fruit fly—despite its apparent simplicity compared with humans—is still an extraordinarily complex organism.

"Biological organisms don't work just one gene at a time," Rubin says. "The genes interact with each other in very complicated pathways and networks. We've tended to oversimplify biology to fit what we could work on in the laboratory. We've had some successes. People have figured out development in the early *Drosophila* embryo, for instance, but that's probably about the most complex thing that we have figured out with this one-gene-at-a-time kind of approach.

"The genome projects are really going to change biology and the way we do experiments. They will allow us to look at the problems in all their complexity and collect all the information we need to approach them at that level."

Each of the model organisms brings

"YOU CAN'T SAY 'I WANT TO CROSS THAT PERSON WITH THAT PERSON AND SEE WHAT THEIR GRANDCHILDREN ARE LIKE.'"

Gary A. Taubes is a California-based science writer.

This fruit fly (enlarged 25 times) grew an extra pair of wings through mutations in bithorax, *a gene that controls development in the rear half of the fly's body.*

advantages the others don't have. A simpler organism like yeast, for instance, is easier to understand, but it is harder to extrapolate that knowledge to the development of humans and human diseases.

For some types of research, it is essential to use multicellular creatures, such as the roundworm or the fly. Yet the roundworm is a pretty simple system, Rubin points out, compared with the fruit fly, which exhibits certain behaviors that can be found in humans and other mammals.

"Things like biological clocks and circadian rhythms, like learning and memory, have all been studied in *Drosophila*," Rubin says. "But you can't study them in a worm because worms are not complex enough. So there are processes that you can study in a fly that you can't study in a worm, and there are things in a worm that you can't study in yeast, and there are things in a mouse you can't study in a fly."

The fly in question, *Drosophila melanogaster,* is about as common a fruit fly as you'll ever come across. While there are some 900 species of *Drosophila* worldwide, *melanogaster* is the one you've probably seen yourself—the tiny flies, maybe an eighth of an inch long, that you'll find hovering around your overripe bananas.

These flies first auditioned as lab animals at the turn of the century, but they were not an immediate success. Biologists and geneticists tended to work on them as a "last resort," after their other breeding experiments had failed, writes University of Pennsylvania historian of science Robert Kohler in *Lords of the Fly*. They wrote to their research colleagues complaining that their laboratories were covered with "bugs, bugs, bugs" or the "pesky little beasts." But these early

Drosophila pioneers persevered, if for no other reason than that fruit flies are ideal for student projects—the colonies are inexpensive to keep and easy to replace if lost to carelessness, an invasion of fly mites, or simply summer vacation.

The fruit fly also has a small genome—an estimated 13,600 genes, compared with the estimated 40,000 genes of humans. It takes only 24 hours for a fruit fly to develop from a single fertilized cell to a larva with a complex musculature, nervous and digestive systems, eyes, and mobility. It takes only 11 days for an egg to develop into an adult female fly that will lay hundreds of eggs. "So it's very easy to generate a lot of offspring in a very short period of time," says Steller. "And you can see the embryos develop, because flies just lay eggs—which is not the case with human and mouse embryos, where the early stages develop inside the mother, in utero, and it's very hard to observe what's going on."

While *Drosophila* research was at the forefront of biology and genetics for the first half of the 20th century, it wasn't until the late 1970s, after the advent of molecular biology, that fly people realized the full extent of what they were learning. "It was then that biologists began to come to grips with the incredible homology across species," says Corey Goodman, an HHMI investigator at the University of California, Berkeley. First came an understanding of how nature lays out the architectural plan of the developing fly embryo. Clusters of genes turn on in sequence, determining the overall body pattern—front and back, for instance, and up and down—and then successive gene cascades lay out increasingly localized structures.

DISCOVERING THE HOMEOBOX

Central to this architectural plan are genes containing the so-called homeobox, a DNA-binding element that was discovered independently in 1983 by Walter Gehring and his colleagues at the University of Basel, Switzerland, and Matthew Scott and Amy Weiner, who were then working with Thomas Kaufman at Indiana University in Bloomington. Homeobox genes encode proteins that tell the cells in the various segments of the developing embryo "what kind of structures to make—antennae for the head, for example, and legs for the three thoracic segments," says Scott. Almost identical homeobox genes were quickly found in the genomes of mice and humans as well.

"Until then," says Scott, "there had been strong sentiment that animals as different as vertebrates and invertebrates would have such different mechanisms of control and growth and patterning that, at most, we could use analogies to compare them, but little else would be directly transferable."

With the knowledge that the same mechanisms were at work, "the whole problem of development was simplified," says Scott. "Instead of having to study genes and proteins and molecular events that are very specific to one tissue, or one developmental stage, or one organism, you can take advantage of the special attributes of whatever your experimental system is, and whatever you learn is likely to be very useful in understanding a great many other events in other tissues, stages, and creatures."

This is what Rubin calls "the take-home message" of *Drosophila* research and the last five years of molecular biology.

Rubin's own work on the development of the *Drosophila* eye has borne

HEADS OR TAILS? FLIES AND MICE SHARE GENES THAT CONTROL BODY PATTERNS

In insects and mammals, very similar genes—in the same order—control the development of front and back parts of the body.

These homeobox-containing genes lie on a single chromosome in the fly (top row of colored squares). Mammals have four similar gene clusters that lie on four separate chromosomes (lower rows of squares).

The genes are color-coded to match the parts of the body in which they are expressed.

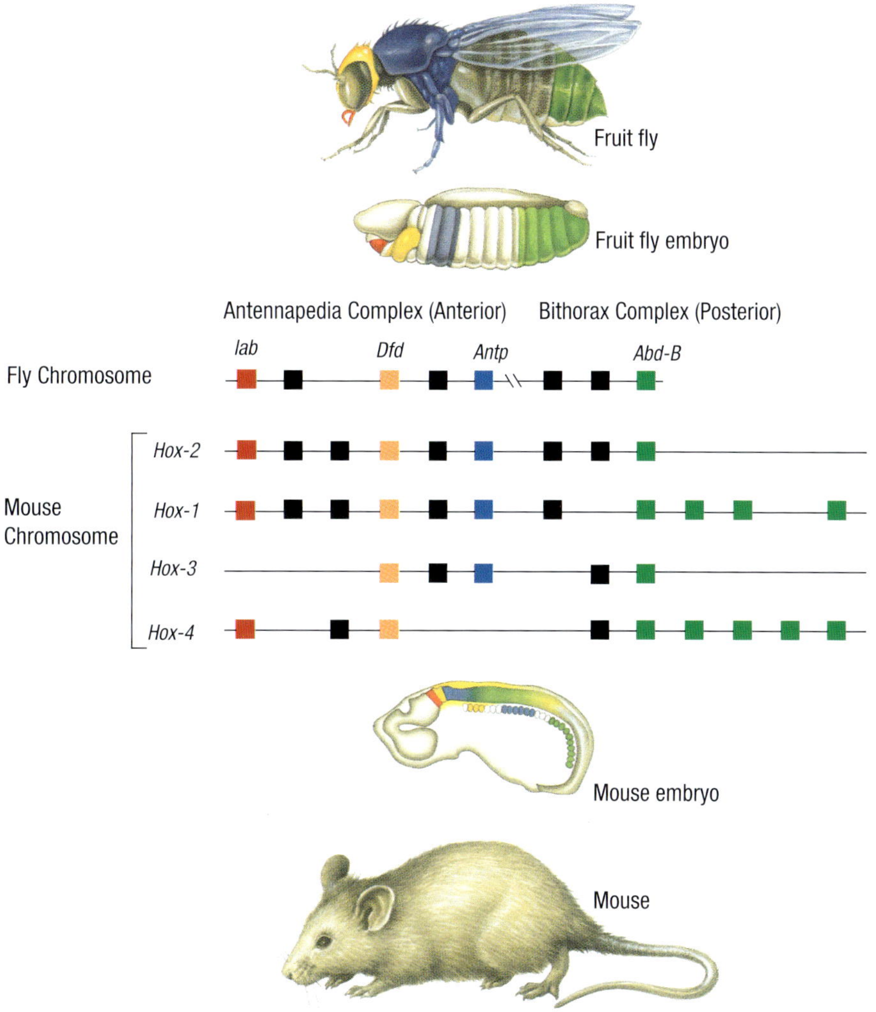

this out. "In the fly retina, for example, we studied a signaling pathway that starts with a receptor called *sevenless*. We worked out this whole, long pathway, with about 20 components identified. Some of those components were already known, but some were new. And in at least two cases, we've identified a new component and then been able to show that it was conserved in humans.

"So in fact we discovered genes that are important for these same signaling events in humans but weren't known before in humans. We were able to do it because the experimental techniques are better in *Drosophila*. We isolate the gene in *Drosophila*, clone it, and then isolate the corresponding gene in humans. That's a very well proven method of gene discovery, and there are now literally hundreds of examples."

In 1986, Scott and his colleagues began studying a *Drosophila* gene known as *patched,* which plays a crucial role in the development of the fly embryo. By 1994, they had discovered *patched* homologs in species ranging from the flower beetle to mice and humans. What's more, they had identified the human homolog of *patched* as the gene that goes awry both in basal cell carcinoma, the most common human cancer, and in a disfiguring genetic disorder known as basal cell nevus syndrome (see p. 41).

In the few years since then, *Drosophila* research has led to the creation of the first animal models of these disorders—mice that develop basal cell nevus syndrome or basal cell carcinoma because of defects in their *patched* genes. Studying these mice should make it easier for scientists to find treatments for the human diseases.

FLIES AND THE HUMAN BRAIN

Even for the most complicated organs, such as the brain, the lessons of *Drosophila* have proved to translate directly to mammals. Goodman, for instance, has been using *Drosophila* to study the wiring of the brain and nervous system. Although flies are not the most intelligent of creatures, his work has already led to a better understanding of how the human brain develops.

"When we started 20 years ago," he says, "we knew almost nothing about how you wire up a brain, especially a human brain, which has trillions of nerve cells, each making hundreds of thousands of connections. Somewhere in the genome is the information that gets read out and somehow allows those cells to find each other.

"I began with the assumption that whatever the fly has, with its brain of 100,000 nerve cells, the human must have as well. If a fruit fly has one cell X hooking up with one cell Y, humans might have 10,000 Xs and 10,000 Ys. If you want to know how to hook them all up in the first place, a good place to start is the fly. I can find out which of the fly's 15,000 genes control wiring in the nervous system. And then, later, I found out that humans did have many of the same genes, used in the same way."

Goodman and his colleagues identified a family of proteins in flies that controls the wiring of the nervous system by chemically repelling the growing nerves away from one cell and nudging them toward another. "Cell X grows toward cell Y," he says, "not just because Y is attracting it but maybe because cells W and Z are sitting there repelling it. So we found this family of repellents, called semaphorins, of which there are 5 members in the fruit fly and maybe 25 in humans, all highly homologous to the ones in the fly."

With so many breakthroughs involving research on fruit flies, *Drosophila* was an ideal candidate for a genome project, says Rubin. Yet at first the Drosophilists saw no need to spend millions on such a grandiose scheme.

"*Drosophila* had been studied for so many years," says Rubin, "that people didn't appreciate the impact a genome project would make. Everyone already had very interesting mutations in flies that controlled interesting things like pattern formation, and they had enough tools that they could actually go isolate and clone those genes. We were moving along at a slow, steady pace, finding out something interesting every six months. It's hard to take people like that and say, 'Now spend

"WE HAVE TO UNDERSTAND WHAT ALL THESE SEQUENCES DO ... AND DEVISE WAYS OF INTERVENING WHEN THEY DON'T WORK CORRECTLY. AT THAT POINT, WE'LL BE PRACTICING MEDICINE."

five years building an infrastructure so you can find something interesting every six weeks.'"

Rubin got involved when his frequent collaborator Allan Spradling, an HHMI investigator at the Carnegie Institution of Washington in Baltimore, visited him in Berkeley in 1991. The two sat around complaining that nobody had jumped in to do a full-scale *Drosophila* genome project.

As Rubin tells it, "We said, 'Someone ought to do it,' and then we realized we were the ones in the best position to do it, and we shouldn't be sitting there complaining and whining why no one else is doing it if we're not doing it ourselves. At that point we thought, 'Let's try to coordinate it and get the project going.'" Spradling and Rubin applied for a grant to do the project at the Lawrence Berkeley National Laboratory, and they recruited promising young scientists to help them. "It all came together within months after Allan and I decided we needed to be more proactive," says Rubin.

There were still skeptics among the Drosophilists who didn't believe the project was necessary. Rubin says they changed their minds when they saw what happened to yeast researchers once the yeast genome was sequenced. "It made them realize how powerful a tool this would be. It's like having a periodic table of the elements. It enables a whole range of technologies. So when people saw what was going on in yeast, they realized they wanted a *Drosophila* genome. The same people who were complaining that we were wasting money were now complaining that it wouldn't be done in one year rather than three."

Rubin and his colleagues originally estimated that they would be able to sequence the entire *Drosophila* genome by 2002. But that feat was accomplished considerably faster through their collaboration with Celera Genomics Corporation, which is run by J. Craig Venter. Venter pioneered the use of a remarkable method of sequencing, known as whole-genome shotgun sequencing, which became more practical thanks to new sequencing machines produced by the PE Corporation, Celera's parent company.

In shotgun sequencing, the entire genome—3 billion base pairs in the case of humans—is randomly broken up into pieces of various sizes. The pieces are then sequenced simultaneously by hundreds of machines working in parallel, and computers figure out how the pieces fit together. "It's a way of getting a lot of information very quickly," says Rubin. Venter wanted to use the *Drosophila* genome as a test case for the human genome, which was his major goal. The fly sequence was published on March 24, 2000 (www.fruitfly.org/sequence/index.html).

"For me, the whole sequence is really just the beginning," says Rubin. "Chemistry didn't end with the periodic table of the elements; biology doesn't end when we have the sequences. We have to understand what all these sequences do—and understand them well enough to devise ways of intervening when they don't work correctly. At that point, we'll be practicing medicine."

A Young Wo

AND A BILLION

man FLIES

Jenica Chekouras, who helped solve a medical mystery, takes a break on the lawn of the University of Wisconsin campus.

There comes a time in the course of growing up when you are confronted with irrefutable evidence that you are simply not like everyone else. For Jenica Chekouras, this first happened when she was six, an only child growing up on a farm in northern Wisconsin, her body freckled with what Jenica now calls "weird moles." Her mother took her to a physician who made the diagnosis that Jenica suffered from Gorlin syndrome, also known as basal cell nevus syndrome, or BCNS, a rare genetic disorder marked by proliferating skin cancers.

Similar tumors occur in the most common kind of cancer, basal cell

UNEXPECTED REWARDS FROM OUR SIMILARITY TO FRUIT FLIES

carcinoma, which affects more than 750,000 Americans. But while many people will someday get one or two tumors of this type, usually caused by too much exposure to the sun, Jenica's body was covered with them.

"The doctors said, 'You have over 200 carcinomas,'" recalls Jenica, who is now 21. "They said, 'What do you want to do with them?' I said, 'I want to take them all off; toss them in the furnace and burn them.' I didn't want to deal with them as a 6-year-old. But the doctors made sure I was well aware that I was going to have to make decisions that most kids don't have to make."

At first Jenica believed her life would not change much. But when she told her first-grade classmates about the disease, "their parents said, 'Don't play with her. Don't have anything to do with her,'" she recalls. "I remember the rather somber day I sat down and counted how many friends I started out with and how many I had left. The first number was 48, the second number was 6."

Over the years that followed, her physicians took to surgically removing Jenica's carcinomas, a few at a time. But there was a futility to what Jenica calls "the classic surgery: scalpel, sutures, the whole nine yards." She hated more than anything having the stitches removed. Then the surgery would leave scars, and the tumors would grow back in any case. And then there were the jaw cysts, another common manifestation of BCNS, which form around baby teeth and, in Jenica's case, grew as large as Ping-Pong balls. The surgeons took those out, along with three of Jenica's teeth and a good portion of her jawbone. But Jenica persevered. She can deal with the disease, she says, and if other people can't, that is their problem. "Yes, I have it," she says. "If you don't like it, there's the door, be my guest."

Now Jenica Chekouras is a senior at the University of Wisconsin–Madison, thinking about going into a career in law unless, of course, she ever finishes that novel she's been working on. Anyone seeing her walk down the street would probably never know she was different. She's a tall, slender woman with an engaging smile, and only a close look might reveal a few spots on her face that look like moles. She's got a boyfriend and not enough time in a busy schedule to visit her doctors. Her dream is to travel around the world and die comfortably in her house when she's 102.

While there is still no real treatment for Jenica's disease, she has played a vital role in solving a remarkable medical mystery. Her contribution has given scientists a new understanding of the genetic basis of basal cell carcinoma—as well as BCNS—and may someday lead to a cure for both.

FRUIT FLIES SUPPLY A CLUE

This story starts not with Jenica but with fruit flies, billions of them. Over the years, these common insects, the length of an "f" on this page, have become an ideal experimental animal for studying how genes control an organism's development from single cell to complex, fully formed creature—a journey that fruit flies make in less than 10 days.

At the University of Colorado, where he set up his first lab in the 1980s, Matthew Scott used *Drosophila melanogaster* to examine two kinds of developmental genes. One kind—the segmentation genes—

divide the fly embryo into segments that eventually become the different body parts of the adult fly. The other kind—the homeobox-containing genes—give these segments their different attributes: a head, two wings, six legs, a thorax, and the 10 sections of the fly abdomen.

"The doctors said, 'You have over 200 carcinomas,'" recalls Jenica.

A segmentation gene called *patched,* which was discovered by the German researchers Christiane Nüsslein-Volhard and Eric Wieschaus, seemed particularly intriguing. Work on it was prompted by the arrival in Scott's lab of Joan Hooper, who is now at the University of Colorado Health Sciences Center in Denver. If *patched* was mutated in a fly embryo, it was as if the compass that determined the polarity of each body segment in the fly was broken. "Structures would be partially reversed," says Scott. "You'd get a mirror image duplication of patterns." In 1989, Hooper and Scott cloned the *patched* gene.

But the researchers still had little idea how the *patched* gene did what it did. To find out, several groups went looking for similar genes in other organisms. It turned out that a fly gene known as *hedgehog,* which was functionally related to *patched,* did have a matching gene, or homolog, in mice. This news made Scott's team even more eager to find mammalian homologs to *patched* itself.

It had proved very difficult to jump directly from the fly *patched* gene to an equivalent gene in mice, however, so Scott and his colleagues decided to use an evolutionary "stepping stone" approach. They started with various insects—first mosquitoes and then butterflies and flower beetles, where in each case they found a homolog. At each step, they learned more about the gene and which parts of it had stayed the same during hundreds of millions of years of evolution.

Finally they felt ready to tackle mammals. After a year and a half of work, a graduate student and a postdoctoral fellow in Scott's lab—Lisa Goodrich and Ron Johnson—found a *patched* homolog in mice. Johnson calls it their "big success" and gives much of the credit to Goodrich, whom he describes as extremely tenacious in her approach.

With the mouse *patched* gene under their belts, Scott's team went after the human version. Finding it turned out to be relatively easy, especially when they looked at the amino acids (building blocks of proteins) for which genes hold the recipes. Each amino acid is specified by a three-base-pair sequence of DNA. When the researchers compared the amino acids made in mouse and human cells, they found that 94 percent of the amino acid sequence of the mouse *patched* gene was identical to that of humans. (In contrast, only 40 percent of the amino acid sequence of the fruit fly *patched* gene was identical to that of humans.)

"We had never worked with any human material before," Scott recalls. But Stanford has a human genome center, so they asked Rick Myers and David Cox, the two geneticists who run it, to map the new gene to a human chromosome for them. "Then it became really interesting," Scott says. "What you get back is this rough map location and a list of other genes in the vicinity, and you look at the other genes and hope something will turn out to be related—a birth defect syndrome, for example."

Starting with the patched *gene in flies, Lisa Goodrich and Matthew Scott found an equivalent gene in mice, which soon led to the equivalent gene in humans.*

Only one disease rang bells in the region near *patched*. That was basal cell nevus syndrome. "The affected tissues matched almost exactly the areas where we knew this gene pathway was working, based on our mouse studies," says Scott. "In people, defects in the *patched* gene can cause spina bifida, polydactyly—which is extra fingers—rib anomalies, and cancer of the skin and of the brain. These are all places where the *patched* gene is active in mice."

Nevertheless, "we had a lot of skepticism in my lab," says Scott. "Rick Myers had already warned me that whenever you map a gene, you always find something near the map location that looks really good—and it's almost always wrong."

Searching through the scientific literature, Scott then located Ervin Epstein, a dermatologist at the University of California, San Francisco, just 40 miles away, and an authority on the genetics of basal cell nevus syndrome. Scott called Epstein, introduced himself, and said, "I think we've got your gene." He launched into an explanation, mentioning *patched* genes, *Drosophila*, and so on.

Epstein now likes to say that his first thought was, What's a *Drosophila?* although when pressed he'll admit he knew. "But I sure as hell didn't know what a *patched* gene was, or anything about developmental biology." Epstein told Scott it was "nice" that he thought they had his gene, and the two agreed to work together to see if it was true.

Epstein, who describes himself as "a dermatologist who sees patients as his day job," had been searching for the genetic basis of BCNS since 1987, when he read a series of papers in the journal *Nature* on a flight from Cleveland to San Francisco. The articles announced the discovery of the gene that causes retinoblastoma, a rare and deadly cancer of the eye that afflicts children. Like BCNS, retinoblastoma comes in two versions. One affects newborn infants and is characterized by multiple tumors. The other hits children when they're older and is usually characterized by only a single tumor.

Back in 1971, Alfred Knudsen, a pediatrician working at the M.D. Anderson Hospital and Tumor Institute in Houston, had proposed an explanation. In the early-onset version of retinoblastoma, he suggested, children inherit a defective gene from one parent. These children are partway to getting the disease the moment they are born. Then, something as simple as an error in DNA replication in a single eye cell, causing a defect in the normal gene that was inherited from the other parent, would send that cell on its way to becoming a tumor.

"If the factory is producing all its cars with defective rear brakes, then 1 out of 1,000 will eventually lose control...."

In contrast, Knudsen speculated, children who develop retinoblastoma later in childhood are probably born with two good copies of the gene but are unlucky enough to have both copies in a cell rendered defective. This would take longer, causing the cancer to show up at a later age.

"Think of it as analogous to losing the brakes on a car," suggests Epstein. "If you have a factory that's producing rear brakes that fail 1 in 1,000 times, and front brakes that fail 1 in 1,000 times, only 1 in 1 million cars will lose their brakes entirely. That's not very many. But if the factory is producing all its cars with defective rear brakes, then 1 out of 1,000 will eventually lose control, and that's a lot of cars."

In 1987, when researchers at the Massachusetts Eye and Ear Infirmary in Boston and the Whitehead Institute for Biomedical Research in Cambridge, Massachusetts, announced that they had identified and cloned the gene that causes retinoblastoma, they also confirmed Knudsen's hypothesis. The gene—known as *RB,* for retinoblastoma—was the first "tumor suppressor" gene ever discovered, so called because cancer develops only when the gene is defective or missing.

The similarity between retinoblastoma and BCNS caught Epstein's attention. "The idea that retinoblastoma comes in two versions—a rare hereditary type, in which there are multiple tumors with an earlier age of onset, and a sporadic type, with single tumors and a slightly later age of onset—threw new light on what happens in basal cell carcinoma," he says. "Basal cell carcinomas more routinely come in people who are not in the first flush of youth, like me, and have been in the sun too much, and they get one or two basal cell carcinomas. But I also knew there was a rare, heritable form called basal cell nevus syndrome, which happens early and leads to multiple carcinomas."

Epstein and his colleagues set about trying to find and clone the gene responsible for BCNS by repeating the steps that the Boston and Cambridge groups had taken in studying retinoblastoma.

First, they obtained blood and tumor samples from as many people with BCNS as they could find by putting notices in medical journals and talking to researchers at meetings, hoping physicians would get the word and send samples from their patients. Then the scientists started painstakingly going through the 23 pairs of chromosomes of each patient, one by one, looking for defects.

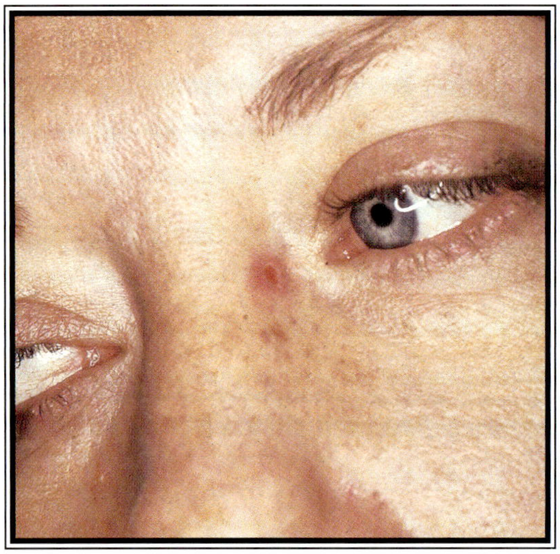

Basal cell carcinomas occur most frequently on exposed areas of the body, such as the face. They may be open sores, reddish patches, shiny bumps, or pink growths with a slightly elevated border and a crusted indentation in the center, such as this one on a patient's nose.

"Tumor suppressor genes are often inactivated by deletions in one copy of the gene," explains Epstein. "Frequently this deletion is not just of a small part of the gene but of a large part of that arm of the chromosome. If you have some way of detecting the two copies of the gene in a normal cell's DNA, you can see if both copies are present in the blood. Then you can look at the tumor and see if one copy of the gene has been lost during its growth."

Epstein and his colleagues started on chromosome 1 because they had seen some hints in the journals that this was the problem area. "Sure enough," says Epstein, "we found some deletions in 25 percent of basal cell carcinomas in area 1q and we said, 'Dynamite, we've put in our thumb and pulled out a plum.' And then we got a very large family from New England with BCNS and looked to see if it mapped there—and it didn't at all! It was clearly not the right site."

Over the next seven years, Epstein and his colleagues made little progress. Luckily, they weren't alone in the search. In 1992, Allen Bale of Yale University, a cancer geneticist, announced that his team had traced the gene responsible for BCNS to an area of chromosome 9.

"Then everyone else found that families with BCNS all linked to the same chromosome," says Epstein. The gene's position was soon narrowed down to a region of chromosome 9q that was a mere 5.5 million base pairs (5.5 megabases) long. As Epstein describes it, it was as though the human genome were the equivalent of the entire *Encyclopaedia Britannica* and they had managed to narrow down the location of the gene first to one volume, and then to one entry. But that entry was still 50 pages long, and the defect in the gene they were looking for could have been the equivalent of a typo in a single word. "And you can't even tell what the words are," he adds, "because they're all run together and there's no punctuation."

This is where Epstein stood in 1995 when he received the phone call from Scott, which he now refers to as his deus ex machina. Soon Scott's and Epstein's groups, working together, proved that the human *patched* gene actually lived in the same 5.5 megabase region of chromosome 9q that had been linked to BCNS and basal cell carcinoma.

Even with that information, it looked as if they might be on the wrong path. When they searched for gross defects in *patched* in the DNA from the 86 families with BCNS that Epstein had collected, "we could find nothing," he recalled. "It was extraordinarily discouraging."

The only thing left to do was look for point mutations in the gene—which, being the equivalent of single typos in the *Encyclopaedia Britannica*, could be time-consuming and maybe futile. Using every trick in the book, the two teams finally did identify some *patched* gene mutations in a handful of patients with BCNS, as well as in a few of the sporadic basal cell tumors. They still needed something that would nail it down, however: a single case, a sample of DNA, or several samples from a family that would be unambiguous. That's where Jenica Chekouras came in.

This abnormality was not present in DNA from either of her parents.

NAILING DOWN THE CONNECTION

Epstein had first met Chekouras in the spring of 1993, when he visited the Medical College of Wisconsin to give a talk on BCNS. Nancy Esterly, a dermatologist at the college who was treating several BCNS patients, including Jenica, invited two of her families in the region down to hear the talk and meet Epstein. "Not that I had anything much to offer them clinically," says Epstein, "but at least I could tell them about what research was going on. So they had come—I don't know—150 miles or something, driving through the mist of early spring in Wisconsin to Milwaukee, and we had a chat. And, of course, they gave us blood as long as they were there.

"We took the blood samples home, and it turned out Jenica had a mutation that was clearly of a type that would disrupt gene function. Her DNA was so abnormal in this

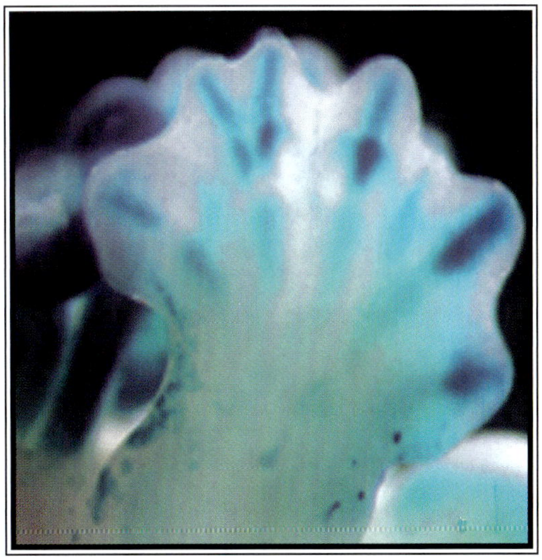

Abnormalities in the first animal model for basal cell carcinoma—a mouse with one defective copy of the patched *gene—include extra digits on its paws. Blue areas show where the gene is active.*

gene that there was no question it would be pathogenic. Most importantly, this abnormality was not present in DNA from either of her parents.

"Since her parents had neither the mutation nor the disease, we could say that the mutation arose in the same generation as the disease, which gave us much more confidence that the mutation in *patched* was really the cause and BCNS the effect. That was the best evidence we had."

Indeed, it was so convincing that Epstein called Chekouras to get a second blood sample from Jenica and her parents just before announcing the discovery of the gene. He wanted to make absolutely sure there was no error. "I was quite surprised to hear his voice," says Jenica. "I'd only met him that once. Now he said, 'I'm sending you these vials by Federal Express and I need you and your parents to give me blood samples and I need them now.' My dad, being an international businessman, was in Italy. So we had to postpone it a couple days until my dad could get back. Dr. Epstein was not pleased. He wanted my dad home now."

Once Jenica's father returned from Italy, the blood samples were taken and sent off, and they confirmed the early result. That clinched it. Epstein, Scott, and their colleagues published their discovery of the genetic basis of BCNS and basal cell carcinoma in the journal *Science* on June 14, 1996. On that same day, Allen Bale and an international team of collaborators announced the discovery in the journal *Cell*. Bale and his coworkers had come to *patched* in the traditional way, beginning with the key segment of chromosome 9q, where they knew the gene for BCNS must be located,

then searching for candidate genes in that region and looking for mutations in patients and families with the syndrome. They were eventually able to identify mutations in the *patched* gene in six BCNS patients, as well as in two basal cell carcinomas not associated with the inherited disease. The two teams shared credit for the discovery.

Scott's group then went about turning a breakthrough in basic research into an important medical tool. The first step was to create a mouse model that could someday be the testing ground for all treatments of BCNS and basal cell carcinoma—a mouse whose *patched* gene did not function. Mice without any working copy of the gene never survived past the embryo stage, proving that *patched* was indeed crucial for development, but mice that had one working copy of the gene (as Jenica does) showed symptoms that were very similar to those of BCNS. Some of these mutant mice were abnormally large, some had extra digits on their paws, and many developed a type of brain tumor, known as medulloblastoma, that results in early death.

Medulloblastoma is the most common malignant brain tumor in children; the mutant mice provided a new tool for studying this dread disease. The mice also developed basal cell carcinomas when they were exposed to ultraviolet light—and became the first animal models for these common cancers.

Researchers now use mice to analyze how such cancers develop and to test potential therapies. Epstein says he can envision the development of a cream that patients could simply rub on their basal cell carcinomas to make them disappear. Several biotech companies are now working on such a therapy.

There are three reasons to be optimistic, Epstein says. "The first is that unlike most cancers, these hardly ever spread to other parts of the body; they do their damage by local invasion. This means the window of opportunity to treat them is much larger than in something like breast cancer, where you have to treat as soon as possible. Secondly, they are superficial and amenable to repeated therapies; you can get at them readily. And then, third, the cells in these skin cancers are already programmed to die. The function of normal epithelium, especially in epidermis, is for cells to divide, differentiate, and die. So there's a reasonable chance that if you nudge these cells over the precipice, they will do the rest of the killing all by themselves."

No one would welcome such a cream more than Jenica Chekouras. Three years ago she started a new form of therapy, called photodynamic therapy, in which a photosensitive agent is spread over her skin and then irradiated with light, which dries up the cancers and causes them to flake off. While the treatment seems to work better than surgery, it's time-consuming and often leaves Jenica with severe acne, painful boils, and burns. "My original idea was to do the photodynamic therapy until every last cancer was gone," she says. "I'm doing what I said I wanted to do when I was six. I'm burning them off. But it's quite physically trying. I'd love it if there was an easier way." ●

Jenica leads a normal life but hopes researchers will find better and less painful treatments for her skin cancers.

UNEXPECTED REWARDS FROM OUR SIMILARITY TO FRUIT FLIES

Ulrike Heberlein observes flies as they go through her inebriometer, a device that measures their reaction to alcohol.

DRUNKEN FLIES

MIMIC HUMAN BEHAVIOR

It was all something of an accident and maybe a bit crazy to boot, says Ulrike Heberlein, a young geneticist who had learned her trade—working with flies—as a postdoctoral fellow in Gerald Rubin's lab at the University of California, Berkeley. She had studied the development of flies' eyes, which was what Rubin did. In 1993, however, Heberlein was already 37 years old and she needed to find a job. "You can only be a postdoc so long," she says. "You have to move on in life and become independent." So she took a job at the Ernest Gallo Clinic and Research Center at the University of California, San Francisco.

The Gallo Center, which is partially funded by winemaker Ernest Gallo, seeks new ways of treating alcoholism. Heberlein's task was to look for the genetic basis of alcoholism in fruit flies.

"They wanted to see if one could use *Drosophila* genetics as a model for alcohol-induced behaviors," says Heberlein. "I thought it was extremely risky. If it didn't work, I would have invested a tremendous amount of my lifetime as a scientist on this. And most everybody else thought it was a bit crazy." But, she adds, "I wanted this job, and I am not necessarily opposed to doing crazy things."

Step one in the research, says Heberlein, was to establish what happens to flies when they become drunk. While the scientific literature, she says, was "pretty vast" on the toxic effects of alcohol on flies, there was precious little on what happened when flies were exposed to low to moderate amounts of alcohol—what happened, in effect, as they went from sober to inebriated.

Heberlein took a few flies, put them in a clear plastic box, and pumped in a bit of ethanol vapor (ethanol is the alcohol in alcoholic beverages). The flies breathed in the vapor and ended up with enough in their systems to make them, by the standards applied to humans, legally intoxicated and a hazard on the highways. The flies' behavior was curiously familiar: "Within the first couple of minutes of alcohol exposure, the flies would become extremely hyperactive," she says. "Soon thereafter they would become uncoordinated as the levels of alcohol in their systems rose, and then they began falling over. Eventually, they just passed out."

Okay, so the behavior of drunken flies was virtually identical to that of other animals, not to mention humans, but that was just the beginning. Now Heberlein went about looking for a clean experimental apparatus that would allow her to quantify the behavior. "That's when we ran into this device called the inebriometer," she says. "It had been used previously in studies of population genetics to selectively breed flies for alcohol resistance."

Regrettably, inebriometers are not the kind of equipment one can simply buy. Even the population geneticists had built their own. So Heberlein and Carol Singh, her research technician and sole collaborator, had to do the same. "The thing with starting something like this," she says, "is the equipment doesn't exist; you have to build it. The first few months you spend going to the glass shop, going through catalogs, finding fittings and flow meters, trying to engineer this inebriometer. I still remember the day we just turned on the air in the first one and we thought it was going to blow up in our faces."

It didn't however, and they eventually built eight of the devices. Each was a glass column, four feet tall and only about three inches in diameter. Inside the column was a series of platforms on which the flies could perch; the columns were attached to the plumbing needed to diffuse alcohol vapor through the column.

"We put flies into the top of the column, where they will stay in the absence of alcohol," explains Heberlein. "In the presence of alcohol, as they start to become intoxicated, they start losing their postural control. They start falling from one of these platforms to the next, to the next, and they gradually tumble down the column. You put 100 flies in the top for each experiment. At the bottom you collect a pile of flies, and that takes a certain amount of time." In total, Heberlein says, she went through something in the neighborhood of "a gazillion flies, tons of them."

She discovered that it would take about 20 minutes for the average fruit fly, known as the "wild type," to end up unconscious at the bottom of the inebriometer. She also tested huge numbers of mutant flies for their ability to handle alcohol, and she found a handful of these mutants whose behavior was distinctly different. Some had become more resistant to alcohol—one mutant, for instance, could survive for almost 35 minutes in the inebriometer before hitting rock bottom—while others seemed to get drunk right away.

Heberlein called this last group *cheapdate*. As it turned out, their mutation was in a gene that had already been identified by researchers at the Massachusetts Institute of Technology (MIT). Known as *amnesiac,* it had been shown to produce memory defects in sober flies when mutated.

The gene produces a brain chemical, a neuropeptide also called amnesiac, that seems to activate a messenger system in fly brain cells. "We don't know exactly how alcohol fits into this," she says. "All we know is that the flies that are more sensitive to the effects of alcohol have a disruption in this messenger system."

Now Heberlein is working to clone several of the other genes she's identified that affect the fly's resistance to alcohol. She is a little reluctant to talk about what she's discovered, but she will admit that she has one mutation, called *lightweight,* that causes flies to be even more susceptible to alcohol than *cheapdate.* So far, no similar genes have been found in humans and no human genes have been conclusively linked to an increased risk of alcohol abuse. But once the human genome and *Drosophila* genome projects are complete, says Heberlein, finding the human versions of these genes, if they exist, should be relatively easy.

With the genes in hand, it might be possible to develop drugs that would modulate their function and control alcohol addiction. "If you absolutely knew what the genes were," she adds, "you could also perhaps identify in advance youths who are at risk for alcohol addiction, and you could employ intervention strategies such as counseling to head off the problem."

Late last year, Heberlein moved from the Gallo Center to UCSF's anatomy department. She is continuing her alcohol work in flies but is extending the approach to a search for the genetic basis of nicotine and cocaine addiction. While Heberlein is not surprised that her crazy project in flies has panned out into a career's worth of work, she is surprised by one simple fact: "I'm still amazed that there are so many similarities between the behavior of these little flies and the behavior of humans when exposed to these drugs."

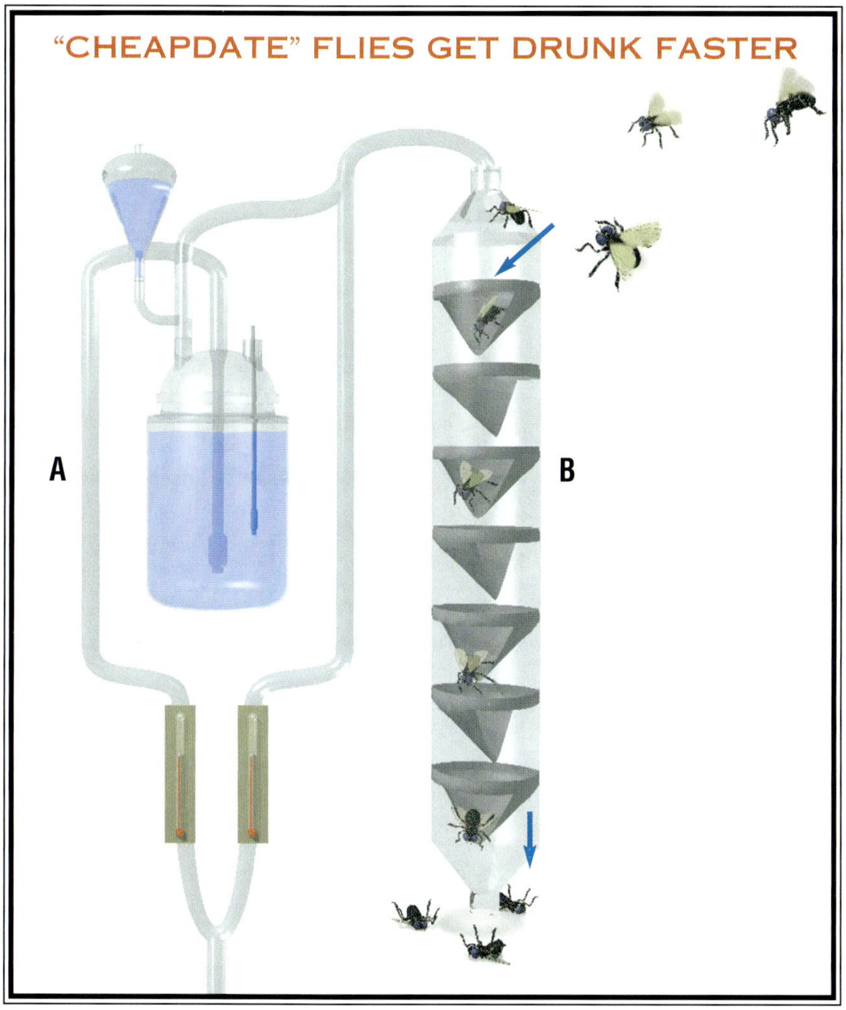

Flies breathe alcohol vapor in Ulrike Heberlein's inebriometer (B), a four-foot-high glass column containing a series of mesh baffles. The vapor is produced by bubbling air through 95 percent ethanol in a tank (A).

At the start of each experiment, 100 fruit flies are placed at the top of the column and exposed to the alcohol-laden air. The longer the flies breathe the alcohol vapor, the harder it is for them to cling to the baffles. After a while, they lose their grip and begin to tumble downward.

It takes normal flies 20 minutes to become so drunk that they fall out of the column. Cheapdate *flies pass out much sooner.*

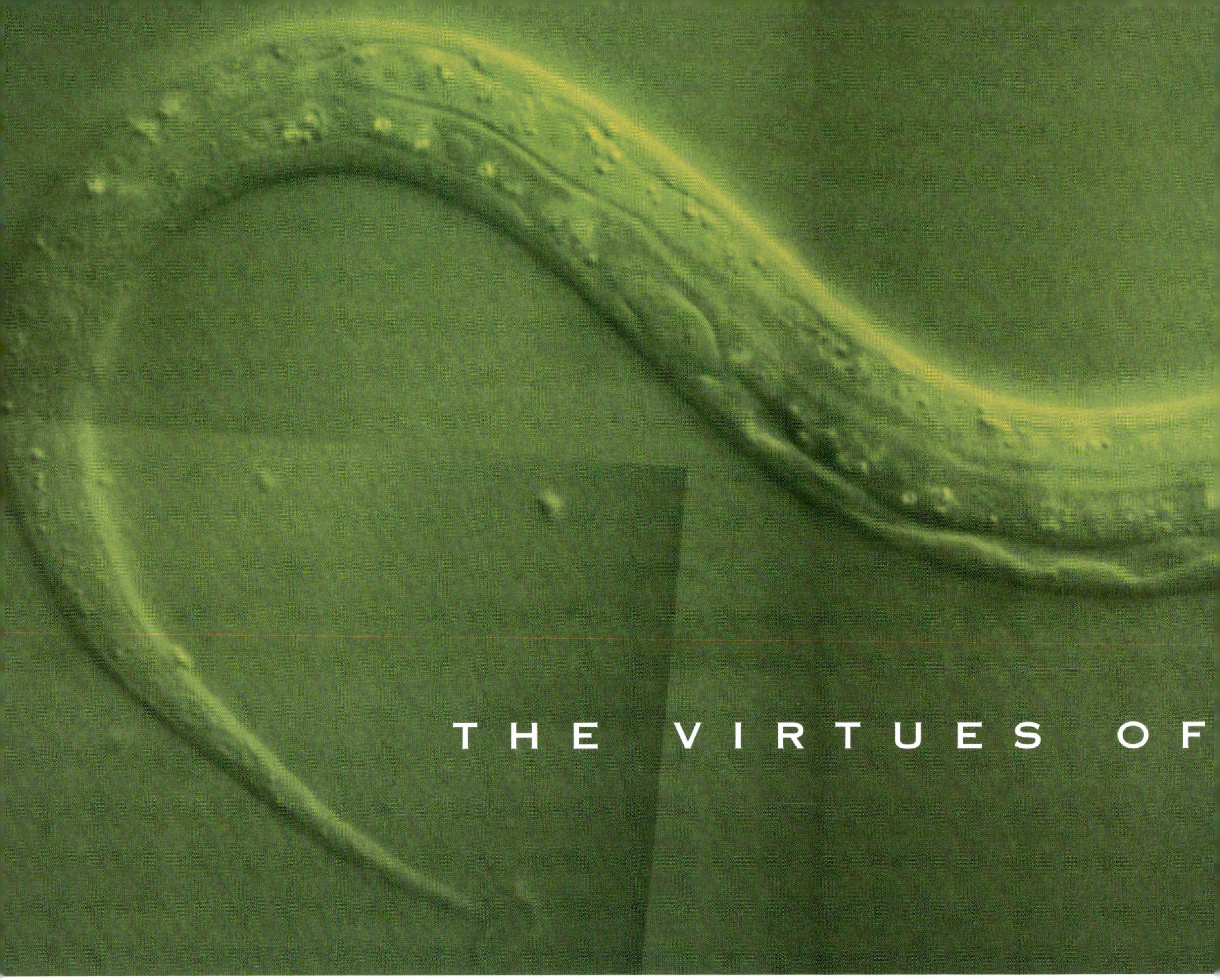

THE VIRTUES OF

A TINY WORM SHOWS HOW

THE FIRST ANIMAL TO HAVE ITS ENTIRE GENOME SEQUENCED: *C. ELEGANS*

By Maya Pines

When Sydney Brenner, one of the founders of molecular biology, chose a worm as a model for studying the nervous system—and, by extension, the human brain—"people thought I was crazy," he recalls. But it proved to be a brilliant choice.

Soon after Brenner began these studies in 1965, a small band of scientists from various parts of the world joined him in his efforts to pry out the secrets of the slithery nematode *Caenorhabditis elegans*. They called themselves "worm people." Their successes—and their numbers—grew. By December 1998 they had succeeded in sequencing the entire genome of *C. elegans,* 97 million base pairs of DNA (genome.wustl.edu/gsc/C_elegans). This was the first time anyone had access to the complete instruction book for making an animal.

CELL SUICIDE

APOPTOSIS BRINGS GOOD HEALTH

This C. elegans *larva will be just one millimeter long when fully grown. Its head is on the right.*

Never mind that the worm they studied is only one millimeter long and almost invisible to the naked eye. What matters is that it is made of hundreds of cells working in unison. Unlike bacteria, yeast, or other single-celled organisms, *C. elegans* has muscles, nerves, sexual organs, and guts. It reacts to touch. It can recognize smells. And it does all this with only 19,099 genes—roughly one-half the number of genes estimated in a human being.

"Why the worm? Why is the Human Genome Project involved in this?" asked Francis Collins, director of the National Human Genome Research Institute, part of the National Institutes of Health (NIH) in Bethesda, Maryland, at the press conference announcing the *C. elegans* sequence. To find clues to the functions of newly sequenced human genes, he explained. "With the lowly roundworm,"

JOHN SULSTON TRACED THE HISTORY OF EACH CELL IN THE DEVELOPING EMBRYO AS THE CELLS GREW, MIGRATED, OR DIVIDED.

Collins said, "we knew we could get the biggest bang for the buck." Both NIH and the Wellcome Trust of London provided financial support for the project.

Work on the worm genome has brought new insights into the mechanisms of Alzheimer disease, stroke, cancer, retinitis pigmentosa (a genetic disorder that causes blindness), diabetes, kidney diseases, and several other disorders.

"The only reason we know anything about what the presenilin genes do in some people who have Alzheimer disease is that we found similar genes in the worm," points out Robert Waterston, director of the Washington University Sequencing Center in St. Louis, which collaborated with the Sanger Centre in Cambridge, England, to sequence the worm's genome.

Early-onset Alzheimer is often caused by changes in a single human gene known as *PS1* (presenilin gene number one). After this gene was isolated biochemically in 1995, it was shown to be similar to four other genes, three of which are also found in *C. elegans*. Researchers at Columbia University in New York City then substituted the human *PS1* gene for a similar gene in *C. elegans* and showed that the human gene works well in the worm. At this news, a large number of researchers in academia and the biotech industry began to study the corresponding worm gene in the hope of finding new therapies for Alzheimer.

The worm people's most important contribution, however, may be their analysis of the curious process of programmed cell death, or apoptosis—the controlled suicide of certain cells for the good of the whole animal.

Apoptosis is essential to all animal life. It is a key part of development and a safety feature in adulthood. As an animal develops from a single fertilized egg, it faces a series of challenges that arise from complexity: It must produce many different types of cells from one cell; it must coordinate these varied cells' activities; and the cells must have ways to communicate with each other. Every cell must work for the common good.

Sometimes this means that a cell must kill itself. Such sacrifices occur in a most orderly fashion. While cells that suffer injuries have messy deaths—they swell up, burst, spill their contents over their neighbors, and produce inflammation—those that commit suicide die neatly and disappear. Theirs are useful deaths, helping to sculpt the body and to keep other cells functioning properly.

For example, human fetuses start out with a duck-like webbing between their fingers and toes; apoptosis makes this webbing dissolve before birth. Only about half of the nerve cells in a newborn's brain manage to establish useful connections to other brain cells; rather than clutter up the developing brain, the remaining nerve cells self-destruct. Millions of immune system cells that might improperly attack the body's own tissue commit suicide every day. All this is for the best—unless apoptosis gets derailed.

Much of what is known about programmed cell death comes from pioneering studies by HHMI investigator Robert Horvitz of the Massachusetts Institute of Technology

(MIT) in Cambridge, who first worked with Brenner in England and later identified many of the genes controlling apoptosis in *C. elegans*.

The worm is particularly well suited to such research. Brenner had originally selected it as a model organism because he wanted a really stripped-down animal, one with a minimum number of cells to investigate. As an added attraction, the worm's cells are clearly visible in its transparent body, especially when researchers use Nomarski differential contrast optics to highlight cell nuclei under the microscope.

John Sulston, until recently director of the Sanger Centre, spent 10 years patiently staring at the worm's cells under a microscope when he collaborated with Brenner. He started with a fertilized egg and traced the history of each cell in the developing embryo as these cells grew, migrated, or divided. Together with Horvitz, he then produced a detailed lineage map of *C. elegans* cells, whose fate was identical in every worm. *(continued on p. 60)*

Robert Horvitz pioneered studies of the "death pathway" through which unwanted cells commit suicide.

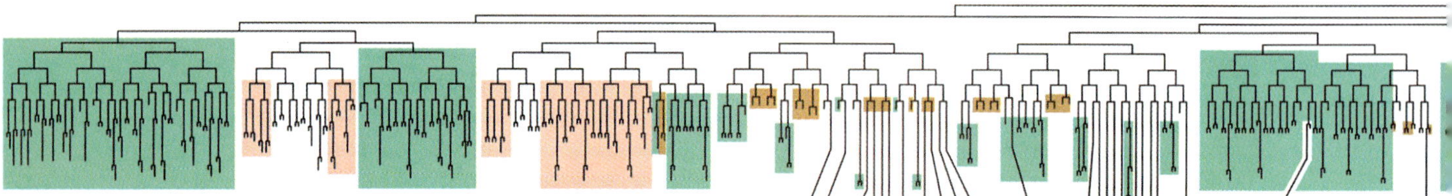

A Family Tree of Every Cell in the Worm

Scientists have learned where each of the 959 cells that make up an adult C. elegans *comes from, tracing it back to a single fertilized egg. As shown on this lineage map, the egg divides into two, and then its daughter cells continue to divide. Each horizontal line represents one round of cell division. The length of each vertical line represents the time between cell divisions, and the end of each vertical line represents one fully differentiated cell.*

Some of these differentiated cells are "born" after only 8 rounds of cell division—for example, some of the cells that generate the cuticle, the animal's coat. Other cuticle cells require as many as 14 rounds. The cells that make up the worm's pharynx, or feeding organ, are born after 9 to 11 rounds of division. Cells in the gonad require up to 17 divisions.

Exactly 302 nerve cells are destined for the worm's nervous system. Exactly 131 cells are programmed to die, mostly within minutes of their birth. The fate of each cell is the same in every C. elegans *nematode, except for the cells that will become egg and sperm. The major organs of the worm are color-coded to match the colors of the corresponding groups of cells on the lineage map.*

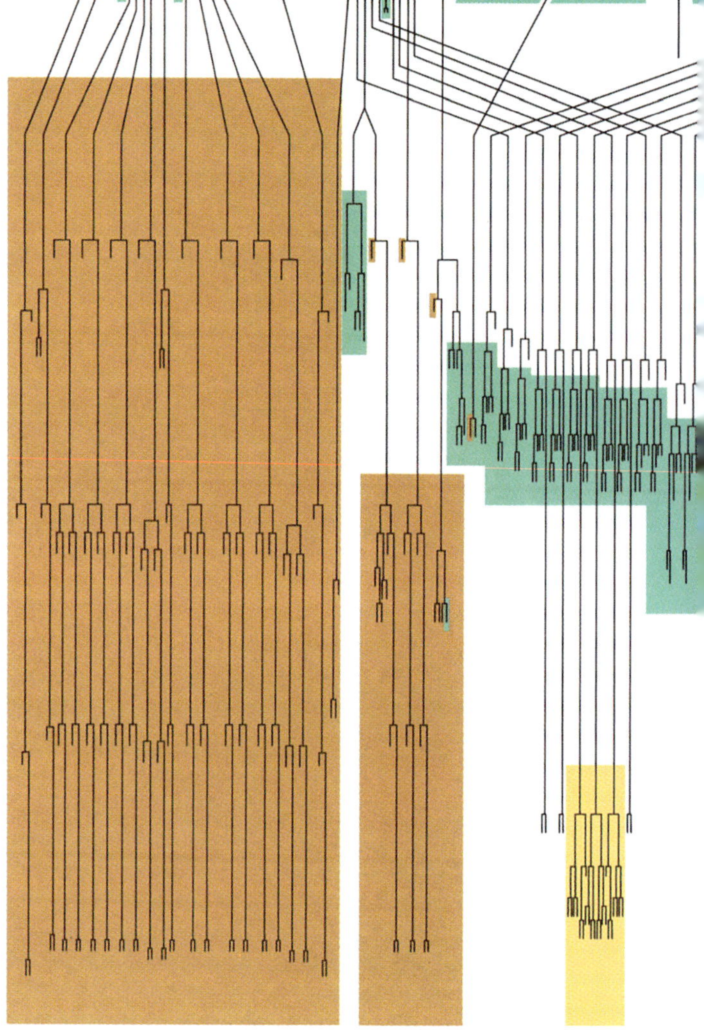

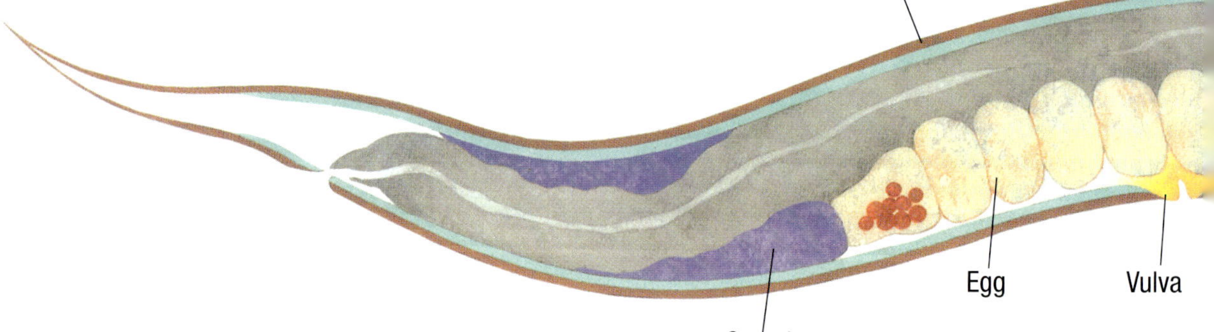

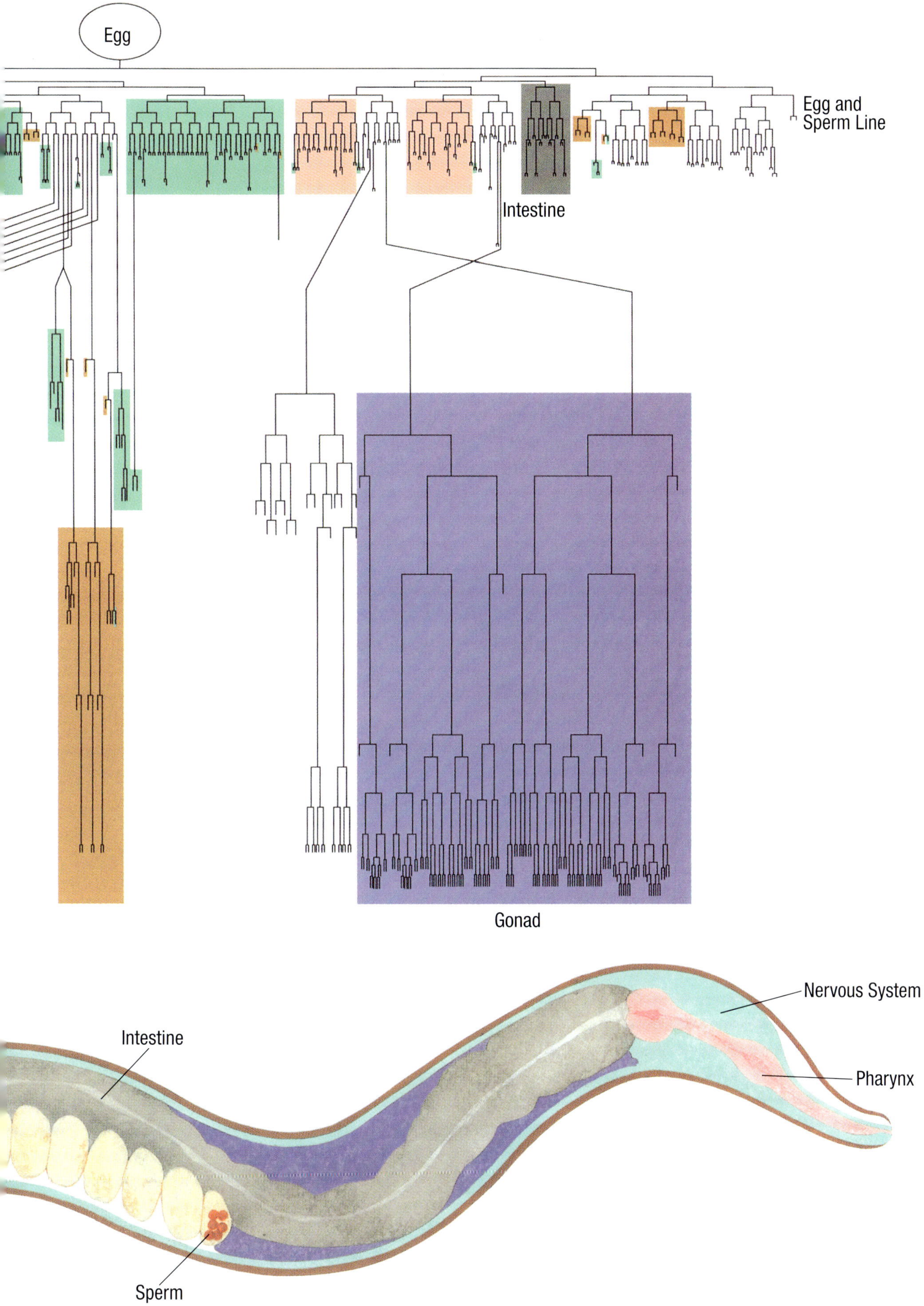
Egg
Egg and Sperm Line
Intestine
Gonad
Nervous System
Intestine
Pharynx
Sperm

A six-week human embryo's finger buds are visible but still webbed. The cells in the duck-like webbing kill themselves about five weeks later, separating the fingers.

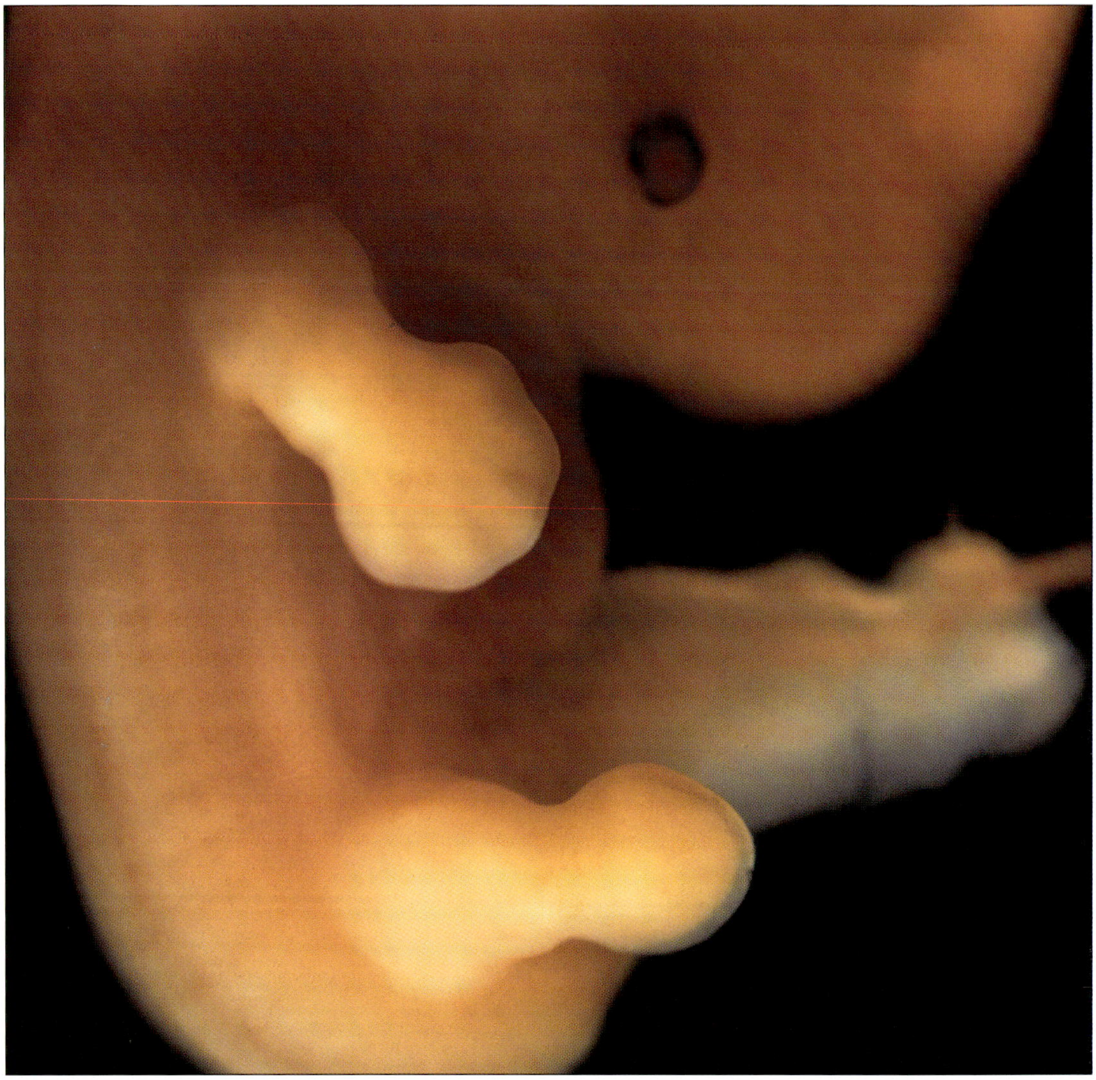

(continued from p. 57) This revealed that although each worm generated 1,090 cells, the adult animal consisted of only 959 cells. Exactly 131 of the worm embryo's cells were programmed to die, often within minutes of their birth.

Focusing on this suicide program, Horvitz identified a "death pathway"—three steps that the doomed cells follow. "First, killing the cell," Horvitz says in a dramatic tone. "Next," and his voice deepens, "getting rid of the body. And third, destroying the evidence."

Basically, he explains, "you produce a corpse and you must do something with it. A neighboring cell swallows the corpse to remove it from the animal. But then the body, now swallowed, must be degraded. This sequence involves a genetic pathway that we have defined in the worm and that may prove universal among organisms, including ourselves."

Each of these steps is controlled by different genes, Horvitz has found. At least 15 worm genes play some role in apoptosis. For example, killing the cell requires the activity of two genes,

"FIRST, KILLING THE CELL. NEXT, GETTING RID OF THE BODY. AND THIRD, DESTROYING THE EVIDENCE."

ced-3 and *ced-4*. "Since these are killer genes," Horvitz explains, "it's very important to regulate them." So another gene, *ced-9,* prevents these two genes from being active at the wrong time and thus protects the worm against unwanted cell death. "*Ced-9* acts genetically upstream of these two killer genes," Horvitz says. A third killer gene, *egl-1,* kills cells by preventing *ced-9* from performing its protective duties.

These findings led to one of the exciting coincidences that occur ever more frequently in molecular genetics. While studying patients with follicular B cell lymphoma, a type of cancer, in 1985, HHMI investigator Stanley Korsmeyer of the Washington University School of Medicine, St. Louis (he is now at the Dana-Farber Cancer Institute in Boston), identified a human oncogene he called *BCL2*. Shortly afterward, Korsmeyer and his associates discovered that *BCL2* leads to cancer not by accelerating cell growth, as other oncogenes do, but by *preventing cell death*—much as *ced-9* does in the worm. Independently, in Australia, David Vaux, Suzanne Cory, and Jerry Adams at the Walter and Eliza Hall Institute of Medical Research came to the same conclusion. Then all eyes turned to the worm.

Human *BCL2* and the worm gene *ced-9* did, in fact, have a similar DNA sequence. Furthermore, when *BCL2* was inserted into mutant worms that lacked a *ced-9* gene, *BCL2* could substitute for the missing worm gene and protect the worm's cells against apoptosis. "This argues very strongly that the worm pathway and the human pathway share components," Horvitz said.

Other similarities followed. The protein made by the worm's *ced-3* (a killer gene) was found to be closely related to the human enzyme ICE (interleukin-1 beta-converting enzyme). A dozen additional human enzymes in the ICE family—caspases, which cut up their target proteins at specific sites—were soon identified. Caspases of various kinds are present in all human cells in an inactive form, but during apoptosis they seem to activate one another in sequence, setting off a storm of protein destruction. Taking due note of these findings, several drug companies have started testing chemicals that might control the activity of caspases, in the hope of blocking apoptosis.

The worm's second killer gene, *ced-4*, also has a counterpart in humans—*APAF1*, a gene recently discovered by HHMI investigator Xiaodong Wang at the University of Texas Southwestern Medical Center at Dallas. Mice lacking this gene die at birth. Many of their brain cells fail to commit suicide during development, and their growing mass eventually takes up so much space that the animals' skulls cannot close around the enlarged brains.

The rest of the death pathway is now being explored, both upstream and downstream of these killer genes. Hermann Steller, an HHMI investigator who recently moved from MIT to The Rockefeller University, is conducting his search in the fruit fly. After looking for mutant flies in which apoptosis fails to occur, Steller and his associates identified two genes—*reaper* and *hid*—that act together to switch on the suicide program in *Drosophila*. Then John Abrams, a former postdoctoral fellow in Steller's lab, identified a third gene, *grim,* after setting up his own lab at the University of Texas Southwestern Medical Center.

"We can take these three *Drosophila* genes and express them in mammalian

A REVEALING LOOK AT PROTEINS IN THE CELL-DEATH PATHWAY OF A WORM EMBRYO

To find out how various proteins in the cell-death pathway of a worm embryo act, MIT scientists gave them fluorescent labels.

The protein made by a killer gene, ced-4, *glows red and is located primarily in the mitochondria of the embryo's cells, outside the cell nucleus. But when the loss of CED-9, another protein in the pathway, triggers programmed cell death, the CED-4 protein moves to the cell nucleus to carry out the sentence. As seen in this picture, the red CED-4 then overlaps with a normally green protein at the nuclear membrane, making it appear yellow.*

This finding suggests that it might become possible to block apoptosis—and diseases caused by excessive cell death—with drugs that prevent the movement of CED-4.

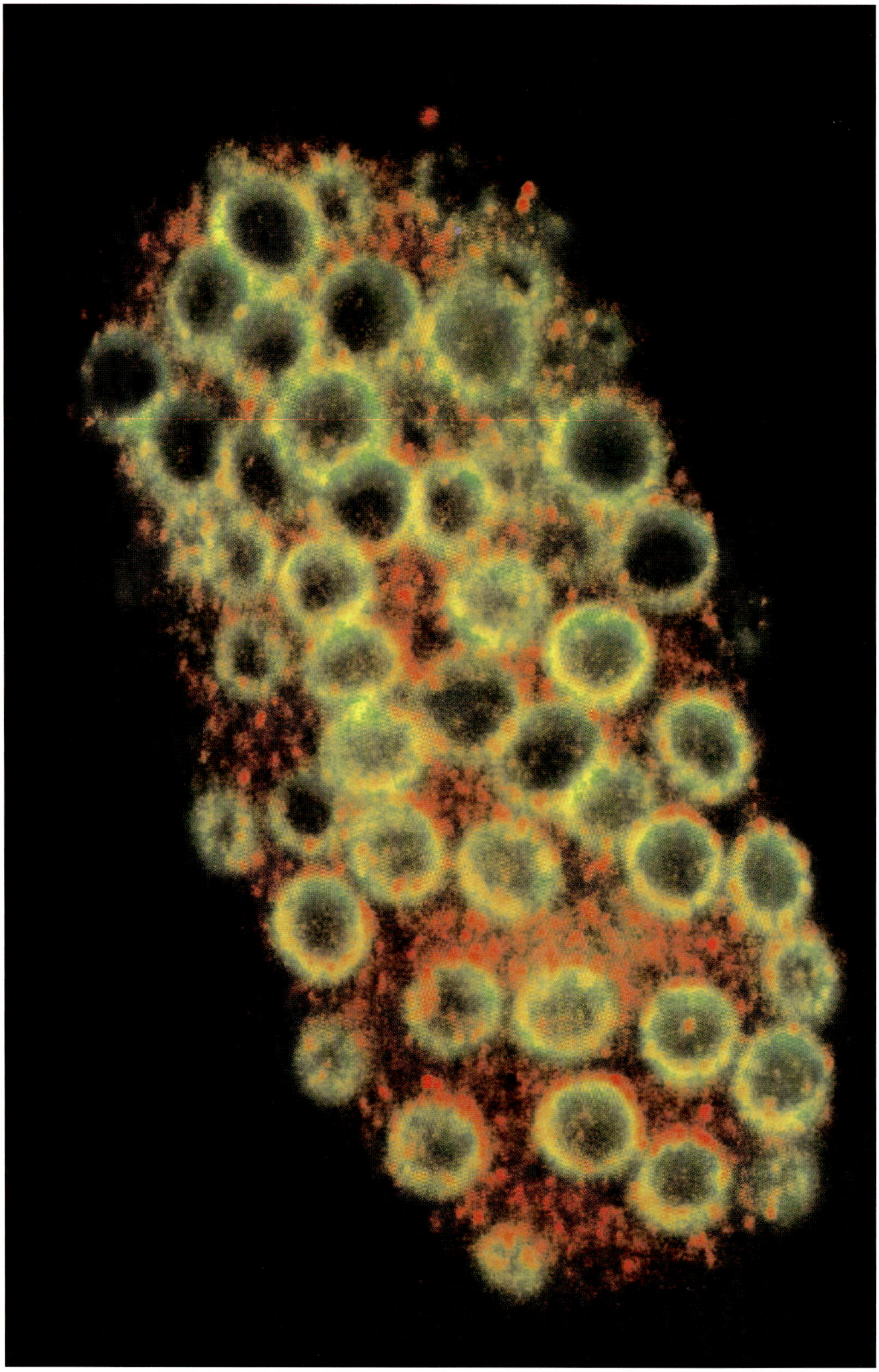

Laboratory of H. Robert Horvitz, MIT

cells, and they induce apoptosis in those cells," says Steller. "So these genes are interacting with other components in the death program in mammals. That makes us absolutely certain there will be mammalian homologs with equivalent function." He adds that the targets of *reaper, hid,* and *grim* are the "inhibitors of apoptosis proteins," or IAPs, which are found all the way up to humans.

Meanwhile, since the development of the fly's eye has been studied in great detail, Steller has been able to look for genes that *modify* the effects of the three death genes. He can identify them simply by examining the eye's condition. Severe defects in the fly's eye, resulting from too many cell deaths, can lead the researchers to genes that promote apoptosis. Healthy fly's eyes can help the researchers identify genes that inhibit cell death. Steller expects that systematic analysis of this sort will lead to a fuller understanding of apoptotic pathways not only in *Drosophila* but also in humans.

He has already developed a fly model of the human genetic disease retinitis pigmentosa, in which retinal cells die, leading to blindness. "The same mutation that causes retinitis pigmentosa in humans causes a very similar disease in flies," Steller says. "So we investigated whether we could block cell death in our fly model—and whether this would lead to retention of vision in these flies. In other words, could we prevent them from going blind?"

He could. "If you block apoptosis in the retina of such mutant flies, the cells will no longer die and, more importantly, the cells will continue to function," he says. "They will not be perfect, but good enough for robust vision." Maybe the diseased retinal cells realize they are not functioning perfectly and therefore kill themselves, Steller speculates. "If you could prevent such cells from killing themselves in a human with retinitis pigmentosa, or perhaps similar degenerative disorders, maybe cells could continue to function and the patient would benefit significantly."

As Horvitz points out, "The number of cells in our bodies is really an

"HOW MIGHT WE STOP SUCH DEATHS? BY BLOCKING THE KILLER GENES THAT ARE RESPONSIBLE."

equilibrium number. Cells are always being added to our bodies, by the process of cell division, but cells are also always being taken away, by the process of programmed cell death. We can generate too many cells—as in cancer—not only by too much cell division but also by too little cell loss.

"On the other hand, the normal pathway that causes cells to die by programmed cell death is sometimes unleashed in nerve cells that were *not* meant to die," Horvitz says. "Some researchers believe that diseases such as Alzheimer disease, Huntington disease, ALS [Lou Gehrig's disease], and Parkinson disease, which are characterized by the deaths of nerve cells, are diseases in which the normal process of apoptosis has gone amok. How might we stop such deaths? By blocking the killer genes that are responsible."

SHOULD WE CARE ABOUT A WORM'S SOCIAL LIFE?

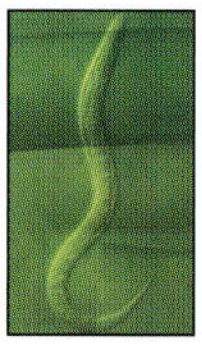

THE FIRST ANIMAL TO HAVE ITS ENTIRE GENOME SEQUENCED: *C. ELEGANS*

A California researcher recently discovered that worms from a particular strain of *Caenorhabditis elegans* don't like to eat alone. When placed near food, they scurry around looking for company. Worms from another strain, however, clearly prefer solitude. She and her colleagues then identified a small genetic mutation that determines this difference.

Another scientist, also in California, noticed that some male roundworms are distinctly deviant in their approach to females. He located a genetic anomaly that accounts for this strange behavior.

You might wonder whether any of this really matters. Should we care about a worm's social life?

Well, yes. After all, we may have similar genes, and even though mutations in these human genes would not be expected to produce the same kind of social isolation or sexual incompetence as in worms, the genetic pathways involved are likely to exist in humans, too. So what we learn from the worm genes may apply to various human disorders.

SOCIAL WORMS VERSUS LONERS

The solitary worms come from a strain of *C. elegans* called N2, whose progenitors were lifted out of a compost heap in Bristol, England, many years ago.

"Just as almost everyone who works in worms can trace his or her lineage back to Sydney Brenner, who started studying them in Cambridge, England," explains Cornelia Bargmann, an HHMI investigator at the University of California, San Francisco, "most experiments on worms are done with the strain of *C. elegans* that Brenner originally started studying—N2.

"These N2 worms are antisocial," Bargmann says. Put them on the kind of food they like best—a patch of bacteria—on the surface of a petri plate, and they will slowly disperse until they have a part of the patch entirely to themselves. Then they stay there alone and munch away. The same is true of about a third of all "wild" strains of *C. elegans*.

By contrast, worms from other wild strains are eager for company. As soon as they sense food, they swarm or clump together, forming eating clubs in certain areas of the bacterial patch (usually around the borders) while leaving the rest of the patch empty. Certain mutant N2 worms also behave in this way.

"These worms like to eat alone," says Cornelia Bargmann, holding up a magnified picture of live but antisocial worms (left). Worms from a different genetic strain keep moving around the bacteria-covered petri plate until they find dinner companions and then clump together for meals (right).

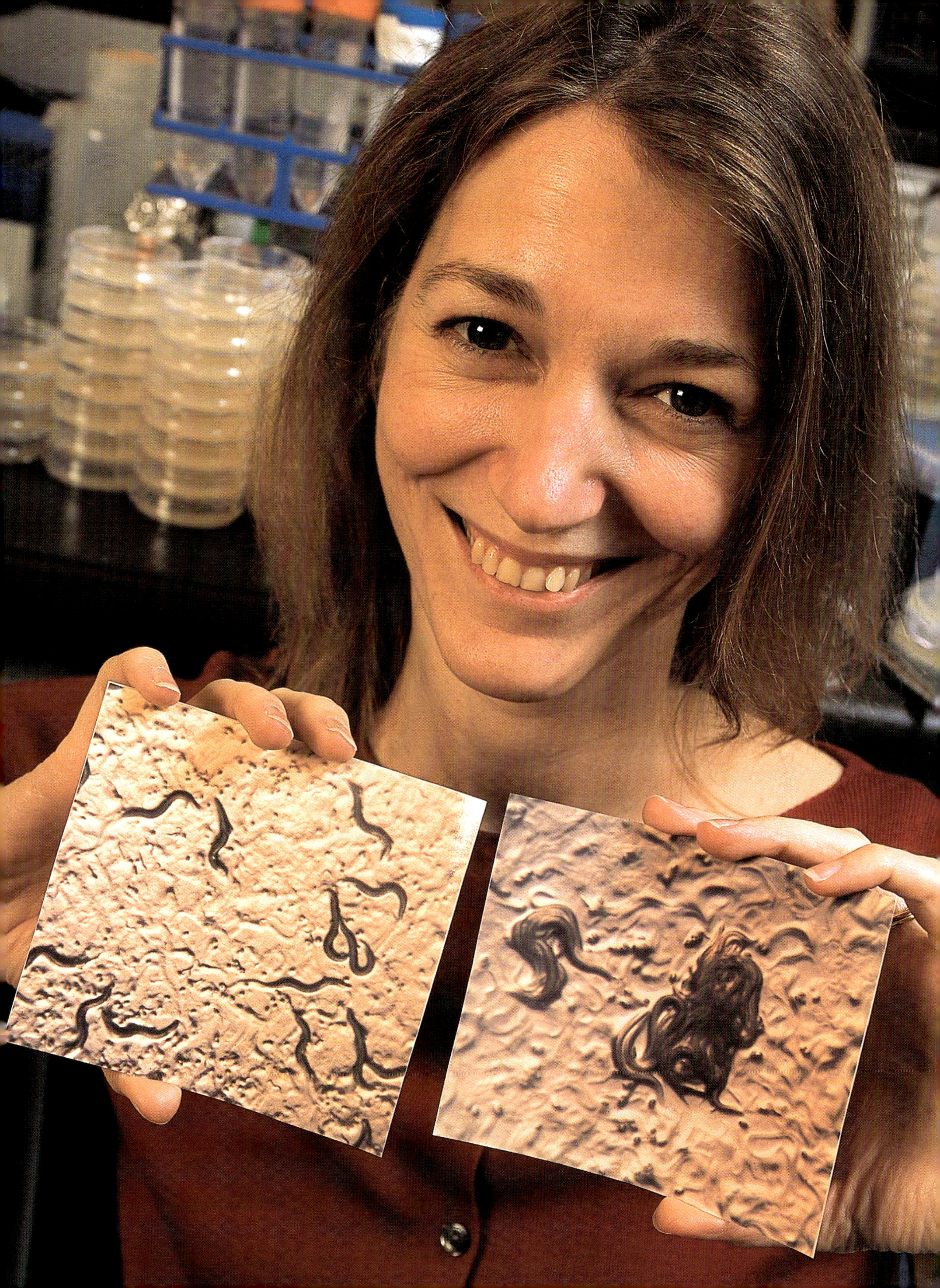

THE SOCIABLE WORMS WERE HYPERACTIVE.

"The original observations about this difference had been kicking around in newsletters and in lore for about 20 years," says Bargmann. A few years ago, she and Mario de Bono, a postdoctoral fellow in her lab, decided the difference was worth investigating. "We became interested in it because we were studying sensory behaviors," Bargmann explains, "and clearly these worms were sensing something about one another, possibly through pheromones."

Behavior is notoriously difficult to pin down, however. "As long as several different wild strains of worms were involved, you couldn't tell whether a particular form of behavior depended on many genes or on a single gene," says Bargmann. "Another difficulty was that the difference between these strains was seen in the behavior of *populations* of worms—not in individuals. You couldn't really score any individual worm for what strain it belonged to; you had to have a whole plateful of worms—at least 50 worms—to get a sense of what the effect was. It's hard enough to score behavior without having to score 50 worms."

Besides, the researchers could never be sure whether a particular behavior resulted from the worms' own traits or whether it was influenced by the behavior of other worms, Bargmann explains. "It's really messy," she says.

Clean experiments became possible when de Bono discovered something that could be measured accurately in a single worm and served as a sign that the worm would search for lunchmates: hyperactivity.

The sociable worms were hyperactive—they kept moving around the food plate rapidly, not slowing down until they found eating companions, he observed. By contrast, the solitary worms moved very slowly all the time, grazing on food right from the start.

When the researchers set about looking for genetic differences between the hyperactive and slow-moving worms, they found a difference in just a few base pairs of DNA that correlated with hyperactivity in all strains of worms. This DNA difference produced a change in a single amino acid. "All the solitary worms—five different strains from Europe and America—have a valine at a particular position in the protein produced by the *npr-1* gene, whereas all the social worms have a phenylanine at the same position," says Bargmann.

De Bono then inserted the solitary worms' version of this gene into a social strain of worms—and the social worms suddenly became solitary. "That says this gene is sufficient for solitary behavior," Bargmann emphasizes.

Next the team discovered that the protein encoded by this gene was remarkably similar to a protein found in human brains, a receptor for a substance called neuropeptide Y.

Biotech companies are extremely interested in neuropeptide Y because it is one of the strongest feeding stimulants found in rodents. Neuropeptide Y is also involved in obesity, anxiety, memory, blood pressure, and digestion. "By studying this protein in worms, we hope to understand how the same molecule functions in mammals," says de Bono. He would like to use the worm to predict how genes regulate blood pressure or eating behavior in humans.

Bargmann sees additional implications of this research. The fact that a sociability trait has now been linked to a gene is "empowering," she says. "We may in fact be finding that genetic influences on behavior are traceable."

Even more important, she believes, this work may open the way to greater knowledge about neuropeptides, "a class of molecules and pathways that we should be looking at more closely," she says, but that in recent years have been "the ugly stepchildren of the neuroscience world" in terms of their popularity and the number of people studying them.

The nervous system has two kinds of transmitters, she explains: the classical neurotransmitters, which send very rapid and precise all-or-none signals between neurons; and the neuropeptides, which modulate neurons over a longer period of time and either increase or decrease the likelihood that the neurons will respond to certain stimuli. Neuropeptides may prove very important in helping researchers understand the complicated genetics of human psychiatric disorders, Bargmann says.

SEXUALLY INCOMPETENT WORMS AND HUMAN KIDNEY DISORDERS

At the California Institute of Technology in Pasadena, HHMI investigator Paul Sternberg recently found another connection between the social behavior of worms and human diseases. Sternberg has a long-standing interest in genes that control the development of sexual organs in *C. elegans*. This subject has brought him many unexpected rewards, from major findings about a cancer gene to the recent discovery of a genetic mutation that holds clues to human kidney disease.

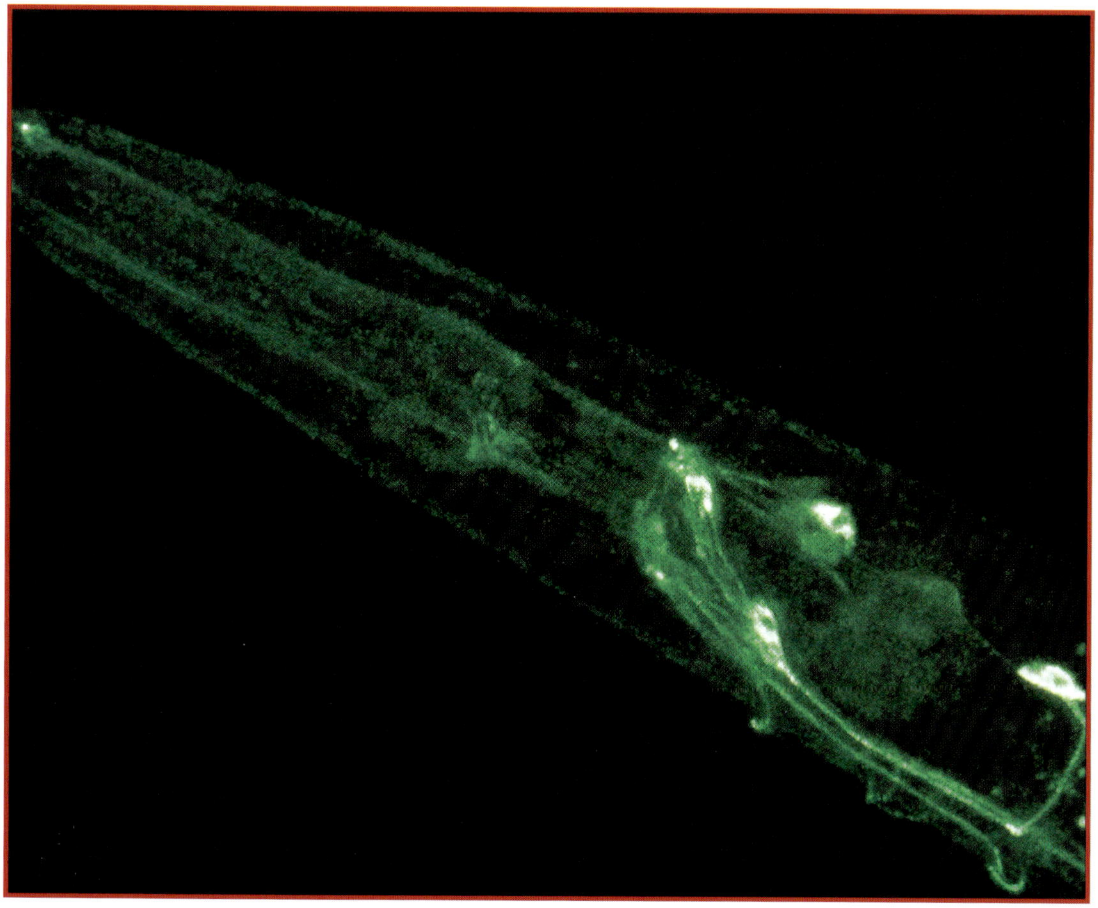

A snippet of DNA in a single gene determines whether a worm becomes a loner or prefers to eat in company. The protein produced by this gene, npr-1, *glows green inside neurons of the worm's transparent head.*

Paul Sternberg sorts slides showing the odd behavior of sexually incompetent worms. Their failure is caused by a mutation in a gene that also exists in humans and is involved in human polycystic kidney disease.

The screen shows a close-up of the male worm's tail with 18 "sensory rays," through which a normal male senses when it has made contact with a hermaphrodite. The male must then locate the hermaphrodite's vulva by using a "hook" (arrow, right).

At first, while working in Robert Horvitz's lab in the 1980s, Sternberg focused on hermaphrodite worms. These worms have a vulva and are really females, but they also produce some sperm, with which they fertilize their first batch of eggs internally. (From then on, they mostly receive sperm from males, thereby gaining additional chances to produce offspring.)

Some mutant hermaphrodites fail to develop a vulva. After self-fertilizing their eggs, they have no way of laying them. The eggs then hatch inside the worms' bodies and the larvae proceed to devour their mothers.

Sternberg and Horvitz discovered three genes that were involved in this error. To their amazement, they found that one of these genes belonged to the notorious *RAS* family of proto-oncogenes—genes which, when mutated, can cause cancer. Mutant *RAS* genes were originally identified in human cancers of the colon, liver, lungs, and bladder, as well as in some leukemias, but researchers had no idea what *RAS* does or what pathway it acts through.

The answers came from the discovery that the *RAS* genes in human tumors contain the same mutation that prevented the worms from developing a vulva. The human *RAS* mutation was also strikingly similar to a mutation that prevented the normal development of the fruit fly eye.

Sternberg then began to study *C. elegans* males, which are considerably smaller than the hermaphrodites. Normally, when a male worm encounters a hermaphrodite, he approaches her, places his tail, which contains sensory structures, flush with the hermaphrodite, and glides along her body until he locates her vulva. Then he inserts two spicules into the vulva to steady himself and transfers his sperm.

While looking for the genes that control each step in this elaborate routine, Sternberg and his colleague Maureen Barr encountered males that acted strangely in the presence of hermaphrodites. Barr noticed that some of these male worms glided right past the hermaphrodites' vulvas, as if unaware of their presence. The researchers then identified the gene that was responsible for this unusual behavior and called it *lov-1* (for location of vulva). They proved its function by showing that if they injected a normal copy of the gene into mutant male worms, the worms found the vulva without difficulty and mated normally.

When the researchers ran the DNA sequence of *lov-1* through various gene databases, however, they could hardly believe what they found: *lov-1* closely matched *PKD1,* a human gene known to be involved in polycystic kidney disease.

"This was a surprise intersection of two different areas," Sternberg says. Though polycystic kidney disease is a fairly common hereditary disorder, it remains poorly understood. Scientists know that a defect in *PKD1* causes the formation of cysts in the kidney and that this potentially fatal condition affects more than 12 million people worldwide. But they have not been able to figure out what *PKD1* does normally, how the mutation leads to cysts, or how the disease might be treated.

"The advantage of working with *C. elegans* is that we can do a lot of experiments much faster than in mammalian cells," Sternberg says. "It would be exciting to find more genes that work with *lov-1* in the worm, since those genes may also work together in the kidney."

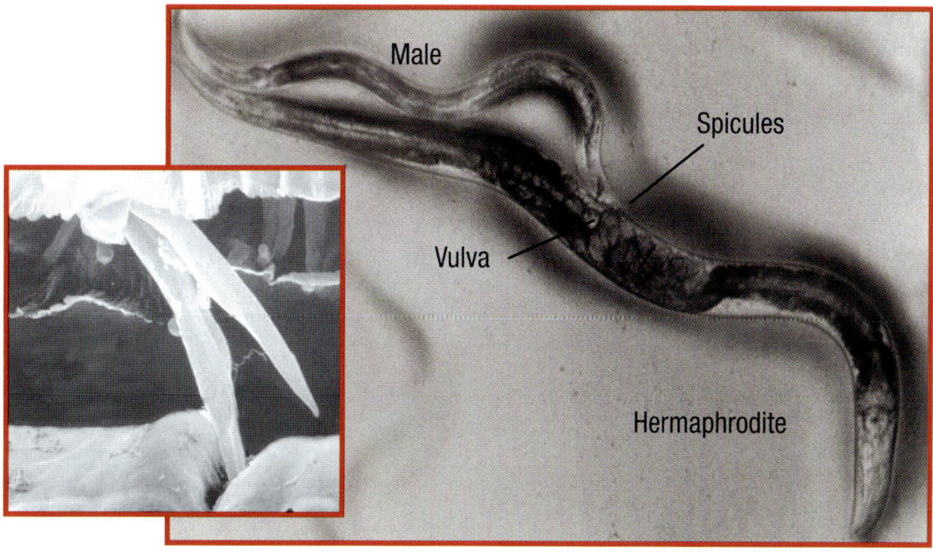

Laboratory of Paul Sternberg, California Institute of Technology

In normal mating, the male stops at the hermaphrodite's vulva and prepares to insert his spicules into it (inset) so he can transfer his sperm.

A LEAP INTO

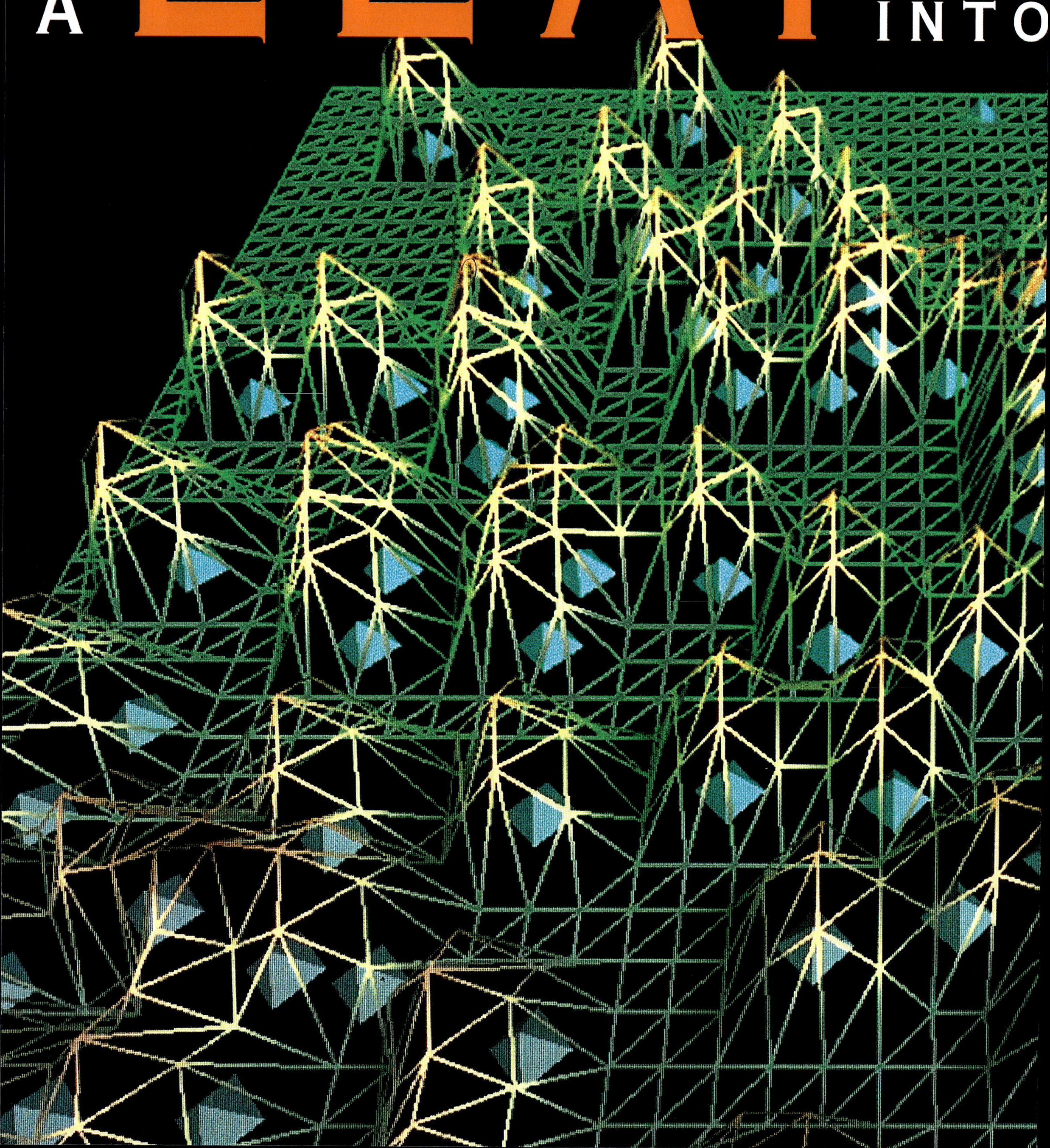

THE WHOLE-GENOME ERA

Researchers known for their work on *C. elegans* frequently get phone calls from scientists who want information about some of the worm's newly sequenced genes. "I am studying disease X and have just cloned a human gene that is involved in this disease," a caller says, "but I have no idea what this gene does. It is very similar to the *C. elegans* gene Y. Can you tell me anything about the function of Y?"

Most of the time, the experts are stumped. Even though the DNA sequences of all the worm's 19,099 genes have been deciphered, only 7 percent of the proteins that these genes encode have been studied at the genetic or biochemical level. At this rate, it might take several decades to learn much about the remaining 93 percent of the proteins in this small creature.

This thought hit Stuart Kim, a professor of developmental biology at Stanford University, with sudden force as the worm genome neared completion in 1997. Kim had been studying the *ras* pathway in *C. elegans*. "It was painstaking," he recalls. "You would start with a completely black box. You

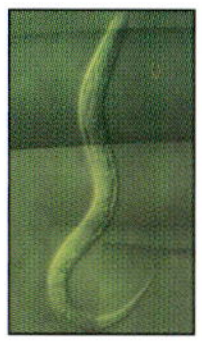

THE FIRST ANIMAL TO HAVE ITS ENTIRE GENOME SEQUENCED: *C. ELEGANS*

Each of the yellowish clumps in this 3-D view of gene activity in a worm cell represents a group of genes that are expressed together and probably have similar functions.

would find one gene and study it in quite excruciating detail, and then after years of work you might actually get to a second gene and then a third gene. That's how we used to build up genetic pathways—step by step. But when you start this way, you never know how big the network is," he says, "and you can't get a really satisfactory answer about how something works until you have the entire pathway."

So in 1997 Kim found himself becoming restless. The worm's DNA sequences had been posted on the Web as soon as they had become available, and most of the genome was known to researchers well before the official announcement of its completion. "I was thinking, the genome is supposed to be so powerful, yet it didn't seem to be affecting what I did," Kim recalls. "It made certain things a bit more efficient, no doubt, but I didn't feel that I was doing anything different from the old stuff."

Meanwhile, a graduate student working with Kim, John Wang, brought exciting reports about the whole-genome studies of yeast in Pat Brown's lab (see p. 6). Wang was friendly with some of Brown's graduate students, who showed him how to do microarrays. "And then, all of a sudden, we were using Pat's equipment to study the worm," Kim recalls. "We used Pat's equipment for two years. And it just worked better and better, and we liked it more and more, and so we got our own equipment. Now we're self-sufficient. In fact, we're helping other labs to do microarrays with worm RNA."

This experience totally changed Kim's approach to research, he says. Since then, everything he has done has involved the whole worm genome.

As he studies the *ras* pathway that leads from a receptor on the cell's surface to DNA in its nucleus, Kim no longer needs to work out, step by step, each of the signals that make a cell grow—or not grow—in response to messages from outside the cell. "With the aid of DNA chips, you can look at all of the cell's genes at once," Kim says. "You can ask, 'When a cell begins to grow, what genes are newly activated inside that cell?' You can illuminate the entire process."

Microarrays will be far more important in the study of worms than of yeast, Kim believes. As a one-celled organism, yeast does not have a nervous system to regulate learning and behavior. Nor does it have any of the major signaling pathways that regulate development. "But animals have conserved signaling pathways that regulate the development of the spinal cord and the brain, how we grow our fingers, how big we make our arms. They regulate—you name it!" he says.

"So that's where I see *C. elegans* as having a profound effect," Kim continues. "We now have DNA chips that contain all of the genes in the worm, and we can get a molecular description of what happens."

In the past, it was difficult to follow all the genes that might be activated at once during cell signaling. "It's very, very complicated to study things that become branched," Kim says. "But the chips get us around that. We can easily identify genes that are turning on in parallel."

Kim is now using DNA chips to study several signaling pathways, including one that directs the development of the worm's germ line (eggs and sperm). "In *C. elegans*, the germ line is more than half of the animal," he says. "The worm is mostly a hollow tube filled with the intestine and the

HE WORM."

gonad [the organ that produces eggs or sperm], which is bigger than the intestine." The rest of the worm consists of a carcass and muscles to make it move, he explains.

Germ-line cells are "immortal and totipotent," Kim points out. "We can all make a baby. We can make everything that's needed for an entire human—the eyes, the nose, the fingers—from the egg and sperm cells in our germ line. And that baby can make another baby." All the other cells in the body, however—the so-called somatic cells—are differentiated into brain, liver, muscle, and so on, and they are mortal. "We will age and die, although our germ line, or the results of our germ line, continue on," Kim says.

What Kim wants to know is how the genes in these two types of cells—the immortal cells in the germ line and the mortal cells in the rest of the body—differ in their activity. With the help of microarrays, he has already identified 1,432 worm genes that are expressed (turned on) in the germ line but not in the soma. "These genes might give us a handle on what makes the germ line so special," Kim says.

Kim's lab also uses microarrays to analyze the process of aging. Some mutant worms live much longer than others, and he would like to know how they do it. "Does *all* aging slow down at the same time in these worms, or just specific kinds of aging in specific tissues? We might be able to get at a subset of genes that are key to making an animal age at the rate it does." This, conceivably, might someday lead to an age-retarding drug.

Such studies could not be done without DNA chips, Kim emphasizes. "It's as if we had a new type of microscope," he says. "We can now use a DNA chip as a gene-expression microscope to look at a process and describe it in molecular terms."

He is so enthusiastic about microarrays that he has volunteered to run such tests for any lab that sends him samples of worm RNA. "We'll just hybridize their RNA and send them back the data," he says. Some scientists have already sent him worm RNA that seems related to a variety of human diseases, such as Alzheimer; others want to find out how nerve cells change as a result of learning. Kim provides such services nationally through grants from Rhône-Poulenc Rorer Gencell, the Merck Genome Research Institute, and the NIH's National Center for Research Resources.

"The grammar used to construct genetic pathways is simpler in worms than in mammals because there are fewer words, fewer synonyms, and simpler sentence structures," Kim wrote recently. "Thus, to learn how to read the book of an organism's genome, it makes sense to start with the worm."

Stuart Kim (left) and John Wang discuss their C. elegans *microarrays, one of which is shown on the monitor behind them. It compares the activity of genes in cells of a male worm and a hermaphrodite.*

The first mammalian gene ever cloned—the mouse gene for beta-globin—is projected behind Shirley Tilghman, a prime mover in the drive to sequence both the mouse and human genomes.

Treasures Await Those Who Study Our Closest Relative Among Model Organisms: The MOUSE

By Stephen S. Hall

The common house mouse has been tagging alongside humankind, mostly as an uninvited guest, since the dawn of civilization some 10,000 years ago. That's about when humans developed the ability to farm, build abodes, store grain and foodstuffs—and spill just enough tidbits and crumbs to support furtive populations of domestic rodents.

"With food reserves in granaries and cupboards, the house mouse began its long interwoven history with humankind," writes Lee M. Silver in his book *Mouse Genetics*. Indeed, he notes that the scientific name of this particular mouse, *Mus musculus,* can be traced back to the ancient Sanskrit word *mush,* meaning "to steal."

Despite that legacy of thievery and stealth, the mouse is about to bestow one of the greatest possible gifts upon human beings: a tool that will help us make biological sense of ourselves. That is because, among all the model organisms used by scientists to explore genetics, the mouse is, biologically speaking, a uniquely kindred spirit. Mice are so genetically close to humans that Ira Herskowitz, a yeast geneticist at the University of California, San Francisco, once remarked, "I don't consider the mouse a model organism. The mouse is just a cuter version of a human, a pocket-sized human."

Over the past 5 or 10 years, biology has seen a dialogue emerge between genetic research on mice

"The mouse is just a cuter version
of a human, a pocket-sized

and attempts at a fuller understanding of human beings (at both the genetic and medical levels). It is a dialogue that grew into a din of conversation as the Human Genome Project, the massive federally financed initiative to map and sequence every human gene, entered its final stages.

A Rosetta Stone

Unlike every other model organism whose genes can be mutated at will, the mouse promises to tell us things about the role of genes in how we think and feel, how we remember, how we reproduce, even how we grow fat. Mice are already telling us how certain human diseases occur, and they will someday tell us how effective new medicines are likely to be. "The mouse sequence will be the Rosetta stone that will help us interpret the human genome," predicts Shirley Tilghman, a Hughes investigator and the Howard A. Prior Professor of the Life Sciences at Princeton University.

"Just look at this," says Tilghman. Spread out upon her lap is a recent journal containing a foldout of human chromosome 22, the first of the 23 human chromosomes to have all its DNA sequenced. "Look at the spaces here and here," she says. A schematic map of the sequence, showing an overlapping series of long, multicolored lanes extending across the page like a psychedelic musical score, indeed possesses numerous unexplained gaps, little patches of terra incognita in the midst of a magnificent new genetic vista just coming into view. "There's so much in this sequence we still don't understand."

Stephen S. Hall, a contributing writer to the New York Times Magazine, *is working on a book about the molecular biology of aging.*

That may be the best explanation for why researchers believe it is essential to keep the mouse as a traveling companion as we embark on our journey into 21st-century biology, even after the human sequence has become available.

"The short answer is because humans are not an experimental organism," Tilghman says. She says it with a laugh, but it's a deceptively simple observation. Just outside her office are rows upon rows of lab benches, outfitted with all the high-powered tools of modern biological inquiry. Here, as in countless other labs throughout the world, researchers are busy erasing, enhancing, turbocharging, and otherwise tinkering with the genes of a mammal biologically close to us, in hopes of creating mutants that display startling effects. Deliberately creating mutants, of course, is not something you can do with humans.

Purely in terms of the logistics of experimentation—of how long it takes to cause a mutation in a gene and then see what happens—mice offer speed and convenience. Females have a three-week gestation period and are fertile again only six weeks after birth, so the generation time can be as brief as nine weeks.

Despite the obvious differences in physical size and appearance, mice are the equal of humans in the only measure that matters in genetics: the size and complexity of their genetic endowment. Like humans, mice are believed to possess approximately 40,000 genes. Like humans, they have a genome that is believed to comprise about 3.2 billion base pairs of DNA. Like humans, they are not only vertebrates but mammals. And as a growing number of cases attest, mice possess a biology and physiology that translate very nicely into their human equivalents and can serve to study human diseases.

Because the evolution of mice diverged from human evolution an estimated 60 million years ago, there are important differences between the species—but even the differences are significant. Mice are evolutionarily close enough to mimic human biology but far enough apart that most of the portions saved, or conserved, must be important. An additional reason why mice—instead of rats or guinea pigs or hamsters—became the mammal of choice lies in the accident of history known as "fancy mice."

A sketch of "Miss Abbie E.C. Lathrop," published in 1913, shows her gazing at one of the fancy mice she bred on her farm as pets.

"Back in the 17th century, in both China and Japan, mice were collected and bred for their beauty," Tilghman explains. Prized mutant mice were bred, much as people breed roses, for strikingly different colors and appearance. Nineteenth-century stock books and catalogs for so-called mouse fanciers list many special strains with ornamental characteristics and names like red cream, ruby-eyed yellow, white English sable, and creamy buff.

"This idea spread to Europe in the 19th century," says Tilghman, "and although the individuals who did the breeding were not geneticists and had no deep appreciation for what they were doing, they were creating inbred lines of mice that were genetically homogenous." That is, hobbyists inadvertently created strains of mice that were genetically identical, and that laid the groundwork for an incredibly powerful scientific approach to understanding biology.

When a Massachusetts woman, Abbie Lathrop, began in 1900 to breed fancy mice as pets on her farm, she also bred mice for researchers at Harvard University. This operation led indirectly to the Jackson Laboratory in Bar Harbor, Maine, which was inaugurated in 1929 with the aim of providing inbred lines of mice—fancy mice and their descendants—to biologists involved in studying cancer and other medical problems.

Even with purebred mouse lines dating back centuries, geneticists needed tools for tinkering with mouse genes. The most powerful of those tools emerged only in the mid-1980s. While working in the lab of Martin Evans at Cambridge University, England, Elizabeth Robertson (now a professor at Harvard University) and Allan Bradley (now director of the Sanger Centre in Cambridge, England) showed that researchers could pluck special cells out of an early mouse embryo—known as embryonic stem, or ES, cells—alter them genetically, and reimplant them. These altered cells then resulted in genetically manipulated offspring.

This first mutation in an ES cell line, Bradley says now, "signaled the emergence of a new age in mouse

It became possible to add or delete specific genes a

Mario Capecchi devised a powerful method of gene targeting that enabled researchers to produce "knockout" mice.

genetics." The trick became figuring out how to alter these cells in a predictable way.

Robertson and Bradley went on to demonstrate in 1986 that retroviruses could be used to create mutations in ES cells. Retroviruses are viruses that have the unique ability to insert their genetic code into the DNA of the cells they infect; thus, genetically modified versions of retroviruses can be used to smuggle desirable information—a new gene, for example—into a cell or to knock out an existing gene by interrupting its coding sequence.

These early mutations were random—researchers couldn't dictate the exact location of the change they wanted to make. Nevertheless, Robertson and Bradley managed to create a mutation in the mouse *Hprt* gene that was similar to the human gene mutation that causes a devastating disorder called Lesch-Nyhan syndrome. It was a sign of things to come.

In 1987, Mario Capecchi, an Italian-born HHMI researcher at the University of Utah, developed a way to create specifically targeted mutations in mice. Researchers could essentially insert or erase a gene of their own choosing in mouse ES cells, using a technique called homologous recombination, and then implant the engineered cells in female mice, which would produce litters of genetically altered offspring.

At times nothing seemed to happen when researchers knocked out a single gene; this was surprising, though useful, information. But in many other cases researchers made spectacular connections between mouse genes and human diseases. The era of the knockout mouse had begun.

Capecchi, for example, discovered that by systematically knocking out genes in a family called *Hox* (originally discovered in fruit flies and often considered master switches that control the formation of the body plan during development), he could produce mice with dramatic developmental defects. When the Utah team mutated two *Hox* genes simultaneously, for example, the forelimbs of mice failed to develop, resulting in animals whose paws grew directly out of their elbows.

In recent years, such tools have become ever more powerful. Bradley is in the process of knocking out mouse genes one by one, as part of

specific locations in mouse DNA.

an international effort to create a vast number of mutant mouse lines that will be made available to all researchers. An increasingly popular technology known as the cre/loxP system has made it possible to control the place and time that a gene is turned on or off in the mouse. Researchers can thus target their mutations to any organ they choose or to any particular time in development.

Among the naturally mutant mice collected over the years at the Jackson Laboratory was a small animal with a brilliant yellow coat known as A^y, or yellow agouti, which had originally been bred in China for mouse fanciers. It turned out to be one of five mutant strains of mice that were obese because of a single gene mutation. In addition to A^y, the other mutant strains were *obese* (*ob*), *diabetes* (*db*), *fat,* and *tubby,* and they proved essential to genetic studies on metabolism and weight maintenance.

"There had been abundant evidence for decades that there was a potent biological system that regulates weight," explains Jeffrey Friedman, a Hughes researcher at The Rockefeller University in New York. During the 1950s and 1960s, all attempts to tease out the mechanics of that system by biochemical means failed, but by the 1980s it became possible to plot an alternative approach using genetics. "By identifying individuals, either human or animal, with a predisposition to disease," says Friedman, "one could go on to identify the gene."

Friedman decided to tackle the issue of obesity around 1986, and the individuals he chose to study were, of course, of the furry, scurrying sort. Both the *ob* (or *obese*) and the *db* (or *diabetes*) mice typically weigh three times what a normal mouse weighs and accumulate five times the normal amount of body fat. During nearly eight years of complicated genetic experimentation, Friedman and his colleagues provided the first glimpse of the biological system that controls the intake of food and the metabolism of fat and showed that the *ob* and *db* mice reflect two aspects of the same biological system.

The researchers discovered that *ob* mice become fat because they lack a single gene, *ob*. This gene, normally active in fat cells, produces a protein that travels through the bloodstream to the brain and tells it, in effect, "I'm full." Friedman named the missing gene's product leptin, after the Greek word for thin (*leptos*). It then turned out that *db* mice lack a gene containing the instructions for making the leptin receptor—the biochemical antenna that receives the leptin signal from the fat cells. In other words, the brain of *db* mice is unable to receive the message "I'm full" dispatched by the fat cells.

Finding these mouse genes enabled researchers to discover the same genes in humans. The initial report on the *obese* gene, which appeared on the cover of the journal *Nature* in December 1994, is framed and displayed prominently on Friedman's office wall (with one unusual annotation: a longtime hockey buff, Friedman asked all-star forward Mark Messier to autograph the journal just beneath a picture showing an obese mouse outweighing two normal mice).

Although the biology of the system is extremely complex and still being worked out, leptin appears to play a critical role in day-to-day biology. "If the fat mass falls," Friedman says,

Several childhood diseases of obesity have been associated

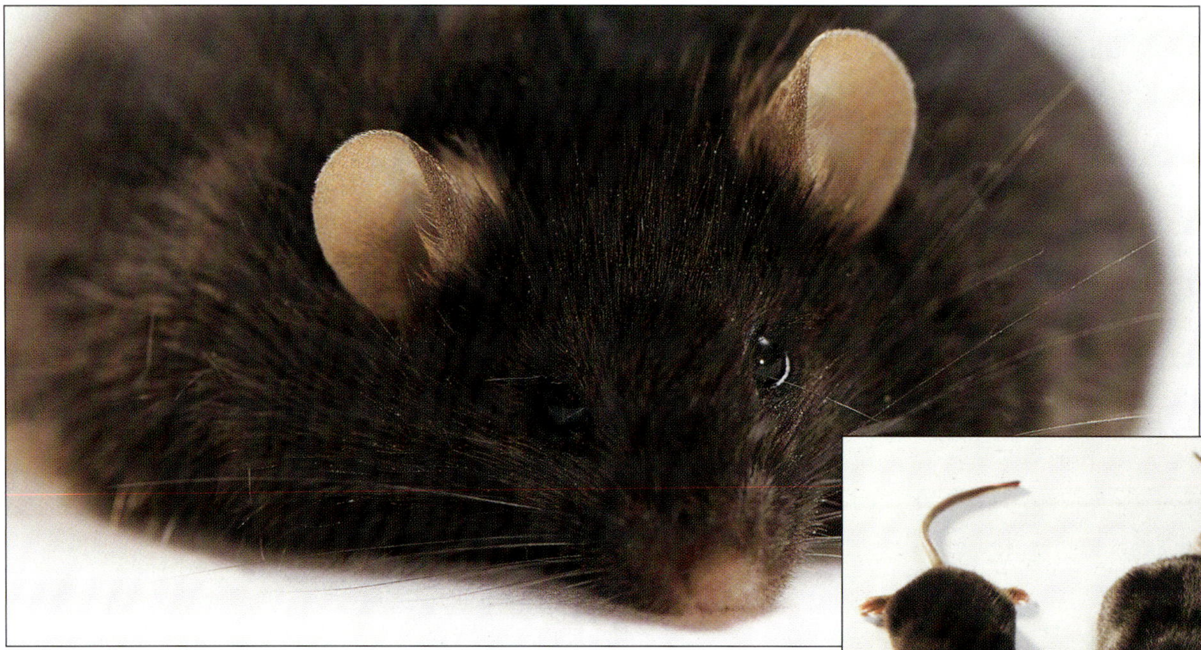

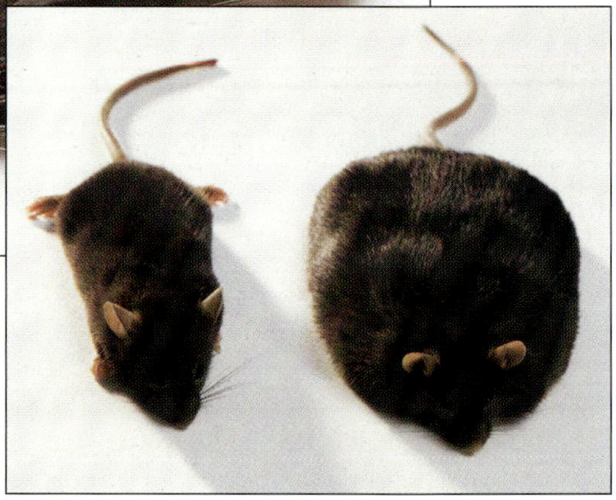

Carrying five times as much fat as its normal sister, this obese mouse lacks a single gene: ob.

"the level of leptin falls, and the urge to eat goes up. After an eating binge, the level of leptin rises, which is a signal to eat less. In addition to modulating food intake and energy expenditure, leptin has an effect on fertility, temperature maintenance, and fat and glucose metabolism." It has become clear, in other words, that leptin sits at the junction of many primal human activities—eating, procreation, basic health, basic activity.

Not surprisingly, leptin appears linked to human disease. Several childhood diseases of obesity have been associated with mutations in leptin genes. Two children in England, for example, experienced horrendous obesity disorders—a two-year-old weighed in excess of 64 pounds and an eight-year-old weighed in excess of 189 pounds. "Both have been treated with leptin," Friedman says, "and both have responded very well."

The leptin story carries within it a seed of caution. Initial press reports about the gene's discovery in 1994 emphasized potential applications for human weight control—an emphasis that the marketplace was quick to embrace. Amgen, a prominent bio-technology company based in Thousand Oaks, California, invested tens of millions dollars to develop leptin as a human therapeutic drug.

In human trials to date, however, the results have been variable: some obese people lost up to 15.6 pounds after six months of treatment, while others did not respond to leptin at all. Friedman points out that many obese people already have high levels of leptin in their blood, which implies they are less sensitive to the hormone than lean people. The reasons for this difference are still obscure. But obesity does sit at the nexus of many basic areas of biology, and tubby mice may yet have a lot to say to the couch potatoes, foodaholics, and compulsive snacksters among us. Friedman's group is currently tracing the influence of leptin as it moves through the brain.

Just Like Cocaine

In 1996, neuroscientist Marc Caron glanced into a cage of genetically engineered mice in an animal room at Duke University and stopped in his tracks. The mice were behaving in a highly agitated manner. Indeed, they were acting, Caron recalls, as if they had been given a maximum dose of cocaine: They were jumpy, slow to calm down, and, as tests would show, about six times as active as normal animals.

"They were wired, so to speak," Caron says, "and we were totally amazed. It was almost beyond anything that somebody could have reasonably imagined."

What made the behavior so striking is that Caron, a Hughes investigator, and his colleagues had simply tinkered with a small housekeeping gene that in essence mops up after the brain's neurons, or nerve cells, have fired. That such a small genetic manipulation could lead to such large behavioral disturbances accounts for the current lure of research on the genetic bases of such fundamental human behaviors as motor activity and memory.

One of the molecules that have stimulated most research in the Caron lab is a blue-collar protein essential to neuronal well-being: the dopamine transporter, or DAT. DAT functions at the very tip of neurons that secrete dopamine, a neurotransmitter that plays a significant role in motor activity, mood, and cognition. When a nerve impulse travels through a nerve cell and arrives at the synapse (the point of contact between adjacent neurons), it causes a quick burst of dopamine to be released; this chemical flows across the synaptic gap and modulates activity in the adjoining neuron.

The DAT protein wheels into action after all the excitement is over. There is always some excess dopamine floating around in the fluid between nerve cells after a synapse fires, and DAT rapidly siphons it up, transport-

Jeffrey Friedman, who discovered the gene that encodes leptin, is trying to trace its effects.

Marc Caron's research on brain chemicals in mice may lead to better treatments for mental disorders.

ing it back into the neuron in a process known as "reuptake." Once back inside the neuron, the recovered dopamine is passed along to another protein, which deposits it in holding tanks called vesicles so that the dopamine can be used again.

"We knew these things must be important," Caron says. But researchers didn't know why, and didn't even have a way to find out until some of the genes for transporter proteins were identified and cloned in the late 1980s and early 1990s. That paved the way for the creation of knockout mice lacking DAT, and the results were, in Caron's words, "quite surprising."

Indeed, these results were impossible to ignore. Mice lacking the ability to recycle dopamine act exactly like animals that have been given high doses of psychoactive drugs, especially cocaine and amphetamines.

Even more surprising, the Caron group discovered in 1999 that mice without the *Dat* gene display several behaviors characteristic of the human condition known as attention deficit hyperactivity disorder, or ADHD, which may affect up to 6 percent of all school-age children. These behaviors include hyperactivity and impaired cognition. The researchers also observed the paradox that, like many ADHD children, the mice respond well to psychostimulant drugs. This calming effect was ultimately traced to another mood-related neurotransmitter, serotonin.

"We've found that transporter proteins play an enormously important role in maintaining the homeostasis of these cells," says Caron. "These mice are extremely hyperactive. They probably represent an interesting model system for neurological and psychological disorders that have been around for a long time, including schizophrenia and ADHD. And the mice become more hyperactive when they are stressed or in a novel situation, which is also reminiscent of the symptoms of ADHD kids."

This research complements what scientists know about a class of psychoactive drugs that affects the reuptake of another important neurotransmitter, serotonin. These drugs, the so-called selective serotonin reuptake inhibitors, or SSRIs, include Prozac and Zoloft. They produce their therapeutic effect by preventing the serotonin transporter protein from effectively mopping up the excess serotonin secreted by a firing nerve cell. In contrast, inhibiting the reuptake of dopamine aggravates behavioral problems.

Caron believes that working on these brain proteins in mice may ultimately lead to treatments for several human diseases that may be caused by disrupted dopamine maintenance, including schizophrenia, Parkinson disease, drug addiction, and an involuntary twitching disorder known as Tourette syndrome.

The notion that serotonin may play a role in the treatment of DAT-related disorders makes these knockout mice extremely valuable as models of human diseases. "Prozac is somewhat of a dirty drug," says Caron. "It increases serotonin in the brain, but serotonin can interact with about 15 different receptor subtypes. So that's a dirty intervention. We think the real potential of using mice here is to decipher which of the serotonin receptors might be involved in attention or hyperactivity in our mice. Then we'll go back and ask, 'Does this apply to humans?'"

One other line of research being done by Caron's group, together with HHMI investigator Robert Lefkowitz, has implications for human medicine—and it, too, began with a strain of knockout mice. In this case, the mice are engineered to lack the gene for a protein called beta-arrestin-2. This is one of several bucket-brigade-style messenger proteins that dampen some of the signals cascading from the surface of a neuron to its center. The protein is activated when a neurotransmitter such as dopamine or serotonin binds with a specialized receptor on the cell surface.

Fiddling with a single component of this cascade, Caron and his colleagues learned, can dramatically increase the effectiveness of opiate painkillers that act through similar receptors. When a standard dose of morphine is administered to mice lacking the *beta-arrestin-2* gene, for example, the analgesic effect is much stronger and lasts much longer. "It really enhanced the potency of morphine severalfold," Caron says, "and it also enhanced the duration of the action." These are important leads toward developing new painkillers and perhaps even better remedies against drug addiction.

Mice and Memory

Another mouse that may someday lead to an interesting drug was recently negotiating a radial maze on the ninth floor of a laboratory at Columbia University, where researchers working with Hughes investigator Eric Kandel have been trying to tease apart the differences between short-term and long-term memory. They do this by testing how mice with specific mutations perform in various mazes.

In this particular case, Gael Malleret, a graduate student, has just placed a small brown mouse at the hub of an eight-armed maze, each arm leading out like spokes from the center and ending in a small dimple in which a sugar pellet had been placed. "The only way for the mouse to remember," Malleret explains, "is to create a spatial map."

"Do you have a smart mouse here?" asks Rusiko Bourtchouladze, one of Malleret's colleagues.

In the age of sophisticated genetic

"So," says Kandel with a laugh,

"if you're a mouse, we car

manipulation, it's a question that has to do with biology as well as learning ability, since researchers have found that laboratory animals' performance in learning and memory can sometimes be altered by modifying a single gene. The mouse the scientists are watching in the maze may be a mutant in which a brain protein has been deleted, but Malleret doesn't know; in blind experiments such as this one, where researchers want to see if a gene deletion affects short-term memory, even the experimenter is kept from knowing whether he or she is testing a mutant or a control animal.

"I'm going to open the gates," Malleret explains, "and the mouse is going to visit each arm of the maze to get the food. So if he visits the same arm twice, it is scored a failure." While a computer tracks the path, speed, and elapsed time of each rodent sortie, the mouse does well for the first five or six arms, but then its memory begins to fail. It revisits two arms previously explored. "Uh oh," says Chris Pittenger, who is also looking on. "*Stupid* mouse!"

For years, the Kandel group has focused on the biological changes that accompany the transition of a short-term memory (one lasting a day or two) to a long-term memory (one that an animal can recall days or weeks later). Kandel shared the 2000 Nobel Prize in Physiology or Medicine for this work. The effort began in a marine mollusk known as *Aplysia,* but the growing ability to modify genes in mice has afforded a look at more complex mammalian biology. In addition to studying mechanisms of memory, the team has started using mice to explore several forms of human mental retardation, including Down syndrome and fragile X syndrome.

In mice, the key event in converting a short-term memory into some kind of permanent record is the synthesis of new proteins, probably as a way of physically altering synapses. The mechanism involves a pair of molecules known as CREB-1 and CREB-2, one of which activates the formation of long-term memory while the other represses it.

Tim Tully and his group at Cold Spring Harbor Laboratory have shown that by increasing the amount of the activator, CREB-1, they could create fruit flies with supermemories in learning tasks; Kandel's group showed that by decreasing the amount of the repressor, CREB-2, they could similarly improve animals' memory. "Tully's studies and our studies," says Kandel, "first made us realize that this protein synthesis step is highly regulated and suggested that defects in the system could suppress memory storage."

Although Kandel began to experiment with mice around 1992, he has kept his eye on human beings as well. "There is very good reason to believe that there is a separate syndrome in aging that reflects itself as a weakening of memory but that can be distinguished from Alzheimer disease," says Kandel. "It's called benign senescent forgetfulness, or age-related memory loss, and it often begins around age 40 or 45. There's a drop in memory, but it's not associated with massive cell loss. So we became interested in seeing if we could characterize this in the mouse."

The mice cooperated fully. Kandel's group discovered that even in rodents, memory loss begins in middle age; only 10–15 percent of 3-month-old mice show memory deficits, while 40 percent of 12-month-olds and 80 percent of 18-month-olds display memory flaws.

The declines in performance parallel a significant drop in an activity called long-term potentiation (LTP) in the hippocampus, a central structure of the brain that appears crucial to long-term memory formation. Mary Elizabeth Bach and Min Zhuo in Kandel's group ultimately linked the memory deficits to decreases in dopamine. Indeed, giving dopamine to the animals reversed the memory loss. "So," says Kandel with a laugh, "if you're a mouse, we can help you."

Dusan Bartsch, a postdoctoral fellow working with Kandel, has extended the mouse work to Down syndrome. Children with this common form of mental retardation inherit an extra copy of chromosome 21, and if they live to the age of 30, they have a greatly increased likelihood of developing early-onset Alzheimer disease. Other researchers had shown that the disease seemed tied to a small piece of the extra chromosome, a piece containing about 50 genes. When studies homed in even closer, it became clear that one gene in particular, the gene for minibrain kinase, was enough to produce a learning deficit in mice.

This small brown mouse in an eight-armed maze may help Eric Kandel identify genes that are involved in mental retardation and the loss of memory.

“We had no idea we were interested in the same gene.”

In talking to parents of children with Down syndrome, Bartsch realized that the retardation seems mild during the first six to eight months of life; therefore, it may not be fully developed at birth. Furthermore, extensive training can sometimes override the defect. So Bartsch created a mouse that models this genetic situation: It overexpresses, or produces unusually large amounts of, minibrain kinase after birth. “When you give mice either of two different spatial tasks,” says Kandel, “it's clear that you get a learning deficit with extra amounts of this one gene, and the deficit can be partially overridden by extensive training.” Bartsch and his colleague Susan Patterson have also shown that these mice have a deficit in LTP, which can be overridden with training.

A Small Window of Time

These mice suggest both that a drug which shuts off the minibrain kinase might be a form of therapy and that there may be a small window of time after birth when such a treatment could be extremely effective.

The availability of knockout mice and other sophisticated molecular tools has transformed the study of memory, says Kandel. “In 10 or 15 years, one problem—the molecular mechanisms through which memories are stored at any given site—may be solved,” he says. “But that will still leave the systems problems. How does the hippocampus interact with the neocortex [the thinking part of the brain] to produce memories, and how do emotions from the amygdala enter in? That's going to interest people well into the next millennium. Not the one we've just begun—the *next* millennium.”

Using knockouts and other clever genetic techniques, researchers have scored one success after another in linking mouse genetics to human health. Elaine Fuchs, a researcher at the University of Chicago, has created a series of mutant mice in which genes related to the protein keratin have been deleted or altered. Almost all these genes involve proteins that are expressed only in the skin and play important roles in the cytoarchitecture of skin cells. Increasingly, these mutant mice are bringing clues to treating various skin diseases in humans.

Sometimes researchers who use mice to investigate unrelated disorders find that their separate discoveries fit extremely well together—to the benefit of all concerned. Several years ago, for instance, Masashi Yanagisawa, an HHMI investigator at the University of Texas Southwestern Medical Center in Dallas, turned to mice to explore some receptors on the surface of endothelial cells that line blood vessels; these receptors may affect blood pressure and hypertension. But when he created knockouts of one of the receptor genes, he ended up with a distinctively pigmented mouse—virtually all white, except for a little tan blaze on the top of its head. He recognized it as a natural mutant that goes by the name of *piebald*—a type of mutant that Shirley Tilghman's group at Princeton used in its research.

At first Yanagisawa could not carry out his research on endothelial biology with these mice because they died within two weeks of birth (the result of an intestinal disorder caused by the same genetic defect). Later on, when he found a way to rescue *piebald* mice in infancy, he

discovered that the adult mutants do have prominent hypertension, as he suspected. But meanwhile he was excited to realize that the gene he had cloned lies at the heart of pigmentation and of a fatal intestinal disorder in humans, Hirschsprung disease, or congenital megacolon, which Tilghman was studying. The knockout mice he had created also failed to develop neurons in the gut—a major feature of the human disease.

Masashi Yanagisawa was surprised to find that he had produced a piebald *mouse (center). The white mouse on the left has a similar, but natural, mutation. The darker mouse (right) carries a milder mutation.*

"Of course, we had no idea we were interested in the same gene," says Tilghman, who has now worked out her own method of keeping the mice healthy until adulthood and analyzes how the disease develops. "It's a great story for how the interplay between two groups can be very synergistic." It's also a great reminder that the conversations between mouse and man, genetically speaking, will be full of illuminating surprises.

Shuttling between Species

to make sense of the human genome

By Maya Pines

On the prowl for genetic similarities that might throw light on newly sequenced regions of the human genome, scientists are taking a fresh look at nearly every form of life, from bacteria to various exotic species of insects, fish, and mammals.

Each researcher has a favorite place to dig in this genetic gold rush—and many are striking rich lodes of information. The newly examined life forms cannot yet match the usefulness of model organisms such as yeast, worms, flies, or mice. But for certain types of experiments, a fresh cast of characters is gaining respect.

The brilliantly colored, 1½-inch-long zebrafish (*Danio rerio*), is one of the rising stars, drawing the attention of more than 1,000 scientists in 250 labs around the world. Tropical fish of this sort are common in home aquariums, but despite pioneering work by researchers in Oregon in the 1980s, zebrafish were largely ignored by biologists until Christiane Nüsslein-Volhard, a tenacious German scientist, put them on the map.

Nüsslein-Volhard had been the driving force behind the systematic search for *Drosophila* mutants (for which she shared the Nobel Prize in Physiology or Medicine in 1995). In the early 1990s, she started an equally ambitious screen for mutant zebrafish. By 1996, her team at the Max Planck Institute in Tübingen and another group headed by Wolfgang Driever and Mark Fishman at Massachusetts General Hospital in Boston had collected more than 1,800 zebrafish mutants with specific defects in blood, kidney, or heart formation.

Zebrafish are vertebrates, like mice, and therefore closer to humans than either flies or worms (zfish.uoregon.edu). Unlike mice, however, zebrafish can be bred and maintained very cheaply. A female

Vertebrates of all kinds fascinate David Kingsley, shown in his office with skeletons of (clockwise from lower left) a rat, an armadillo, a chicken, a pigeon, a frog, and a salamander.

zebrafish will typically produce hundreds of eggs at a single laying. These eggs are easily fertilized. The resulting embryos, which are translucent, grow outside the mother's body, so mutants can be observed at every stage of development.

"When you work with other organisms, you are looking at snapshots," remarks a zebrafish researcher. "With zebrafish, you can see development as a real-time process."

Impressed with these advantages, Leonard Zon, a Hughes investigator who was studying blood disorders at Children's Hospital in Boston, asked Nüsslein-Volhard if she would give him some mutant zebrafish. She obliged, providing 39 mutant embryos that failed to form red blood cells normally; as could be seen under the microscope, their red cells were either too pale or too few.

Zon's team mated the mutant zebrafish to each other and then analyzed their offspring. In short order, the researchers found fish equivalents of three human blood diseases: porphyria, a genetic condition in which red blood cells burst open, causing a severe anemia; another form of congenital anemia; and thalassemia, which results from defects in the production of hemoglobin.

In fact, six different zebrafish strains were found to have some equivalent of human thalassemia, "but only one of these mutants is linked to globin [the component of hemoglobin that accounts for the color of red blood cells]," Zon says. "So there must be five other genes that regulate globin expression, yet nobody knew about them." He hopes to "go on a fishing expedition" soon to isolate these and other "novel" genes.

Even though fish have cold blood and human blood is warm, the two kinds of blood are "very, very similar," he says. "The program is highly conserved."

So, apparently, are the programs for growing a heart, kidney, and other organs. Mutant fish embryos have already provided examples of 100 abnormalities in heart development. In the cardiovascular research center at Massachusetts General Hospital, for instance, Mark Fishman is studying a zebrafish embryo that was named "Santa" for its big heart, and another one, "Slo Mo," whose

A rising star among model organisms, the 1½-inch-long zebrafish is helping researchers find genes involved in human blood and heart diseases.

heart beats too slowly. Both defects also occur in humans, but their causes are more likely to be tracked down in zebrafish.

Zon, meanwhile, has prepared a map of known zebrafish genes that can be compared with genetic maps of mammals, including humans. He wants to promote what he calls "genomic ping-ponging"—the increasingly popular process of going back and forth between different species, using the information from one to fill gaps in the other.

FOCUSING ON SKELETONS

How do biologists choose the animal model or lab samples they will spend so many hours studying? Some researchers start with a problem that they find intriguing and then look for suitable models, while others become enamored of a particular organism and use it exclusively. David Kingsley, an HHMI investigator at Stanford University, was always fascinated by how organisms develop and by evolution, but he also wanted to make sure his work would be directly relevant to human disease. So he decided to focus on something "unique to vertebrates"—vertebrae, the bones of the spinal column. And he chose to study skeletons.

The embryos, which are translucent, grow outside the mother's body.

Leonard Zon points to a zebrafish mutant swimming in a tank in his lab. As he explains to Eric Liao, a student researcher, the fish was found to have the equivalent of thalassemia, an inherited blood disorder.

The mouse was the obvious place to start. Looking through a mouse genome map in 1990, Kingsley put a red dot over each place where a genetic trait that affected some aspect of skeletal development had been mapped to a mouse chromosome. He recalls that "there were 150 different mouse genes that affect, or disrupt, various aspects of building the skeleton. They affect bone formation, bone repair, how many bones you have in your finger, how well you can replace bone fractures, whether you get arthritis, what your tail or vertebral column looks like—tons of different skeletal traits."

His research team then went through that collection systematically to try to isolate genes that control key steps in skeletal development. "It was almost a five-year project at that time to pick one red dot and say, 'This looks like a really interesting gene. It maps to this region of a chromosome. Let's try to isolate all of the DNA that's in that region of the chromosome, decode all the genetic information that's in there, and find the gene that's responsible for this particular trait,'" Kingsley recalls. "The genome project is dramatically shortening the time it takes to find such genes and their pathways."

Working with mice in this collection, Kingsley's team was able to identify genes that control where and when bones form in the skeleton, as well as where joints form between various skeletal elements.

The team also studied skeletal diseases. "Some of the old classical mouse mutations control traits like arthritis," Kinglsey says. "We have recently isolated one of these genes and shown it encodes a completely novel protein that controls mineral deposition in joints. This gene is highly conserved in vertebrates, but it's not in the worm or fly databases, and we haven't been able to find a copy of it in any invertebrate." Not everything is conserved across different animals, Kinglsey points out. "We wouldn't have identified this gene without studying vertebrates themselves."

"Crossing a bat and a horse is not something we're stupid enough to try."

As his work proceeded, however, Kingsley began to marvel not only at the large number of similarities between the basic skeletal structures of various kinds of vertebrates but also at the structures' diversity. His lab is filled with animal skeletons of differing sizes and shapes.

"Look at all these different mammals," he says, pointing to the skeletons of a rat, an armadillo, a chicken, a pigeon, a frog, and a salamander. "They are all made of the same cartilage and bone, but each organism is able to take that tissue and mold it into a different shape to produce an organism that is adapted to a particular function.

"If you want to make an organism that can fly, like a bat, you have to take what used to be hands and somehow stretch out all of the fingers in just the forelimbs—not the hindlimbs, because you still want a traditional grasping foot down there—to form a skeletal structure that can support an adaptation to flight. If you want to make an organism that can swim, look at a seal: it has many more bones in the hindlimbs in order to produce a flat, flipper-like structure that serves to paddle the water. If you want to make an animal that can run fast, you get rid of lots of bones. You don't want all this stuff flapping around; you want a couple of long things that you can run on efficiently.

"So the limbs of a bat, a seal, and a horse look very different from each other. And we still have no idea what types of genes and molecular changes are responsible for these dramatic differences." Normally, when geneticists want to understand the basis of differences between two organisms, they cross the organisms and map the genes that are responsible for the corresponding traits in their offspring, Kingsley explains. Unless the two organisms belong to the same species, however, they produce no offspring.

"Crossing a bat and a horse is not something we're stupid enough to try," he says. "But conceptually, that's the kind of thing we'd like to do. So we tried to think of a realistic way of doing crosses between species that have different skeletal shapes."

The solution that Kinglsey and HHMI associate Katie Peichel came up with was to focus on organisms that had diverged only recently in evolution—so recently that any barriers to mating could be overcome by artificial fertilization.

After a long search, they settled on a little fish called the three-spine stickleback, which boasts a fearsome armor made of bony plates, much like an armadillo's. The three-spine stickleback needs this armor as protection against predators in the ocean, where it normally lives. But it often breeds in freshwater streams and lakes, as close as possible to the North Pole because it likes the cold.

As the glaciers melted at the end of the last ice age, thousands of new streams and lakes were created along the coasts of North America, Europe, and Asia. Many of these were colonized by three-spine sticklebacks, who then encountered new kinds of

David Kingsley has identified several mouse genes whose mutations cause skeletal defects, including Gdf5, *a gene required for joint development. The protein produced by this gene glows blue in the mouse embryo shown on the screen beside him.*

food and predators in each lake. "In the 10,000 years or so since then, those sticklebacks underwent a dramatic adaptation to whatever local environment was in the lake," Kinglsey says.

It's a very recent event, he emphasizes. "The ancestral population of sticklebacks is still swimming out in the ocean with its three prominent spines on the back, but in the different lakes it changed into all kinds of shapes: giant sticklebacks, pygmy sticklebacks, two-spined sticklebacks, zero-spined sticklebacks, fully armored sticklebacks, partially armored, and no-armored sticklebacks. It's such a dramatic variation in skeletal structure, in fact, that when naturalists first discovered all these fish, they said, 'This looks like 40 different species of fish,' because of how obvious the skeletal changes were."

Later on, Kingsley continues, scientists realized that all these fish were still closely enough related to produce fertile offspring. Some of the different-looking sticklebacks would not breed with each other naturally, but they could be bred together by artificial fertilization. "There is no genetic incompatibility among these fish," he explains. "If you collect sperm and eggs from sticklebacks that look very different from each other and do artificial fertilization, you get completely viable, fertile hybrids. Then you can do a thorough genetic analysis on them."

He adds that if one tried to do artificial insemination with a bat and a horse, for instance, it wouldn't work at all. "They are too far apart from each other," he says. "The divergence time is too great. It isn't 10,000 years ago but about 65 *million* years ago that mammals diverged—after the meteor hit the Earth. That's an enormous difference.

"But it's still possible to take a formal genetic approach to working on the differences between sticklebacks, even though the skeletal differences are often great. I mean, you're taking what could be 40 armor plates and turning it into zero armor plates; it's as big a difference as many of the shape variations you'd see between different mammals."

"Each lake is its own independent evolutionary experiment...."

Kingsley's team still works mostly on mouse genetics. But Kingsley and Katie Peichel have made several trips to lakes in British Columbia, Washington, and northern California to collect samples of interesting sticklebacks. They raise these fish in tanks built in a special cold room in the lab.

"Here's a pretty example," says Kingsley, pointing to a tank. "See that guy back there with the red tummy and blue sides? That's a male in breeding colors. There has actually been a huge amount of *behavioral* work done on sticklebacks, particularly by Nikko Tinbergen, the Danish ecologist who won a Nobel Prize for it in 1973.

The bright red underside of the male stickleback (left) shows his eagerness to mate as he approaches a female stickleback with three prominent spines and an alluringly swollen shape.

"Sticklebacks have an elaborate breeding ritual, and Tinbergen worked out a series of visual cues that trigger each step. Red bellies in males, swollen shapes in females, zigzag dances, and trips to nests all serve as triggers for the next step in the overall mating behavior." Because of this interest, which started 40 to 50 years ago, many scientists have studied stickleback ecology, evolution, and behavior, producing a bibliography of more than 2,000 papers.

"We are trying to bring a new, molecular-genetic approach to this organism," says Kingsley. "We have already generated more than a thousand genetic markers and constructed the first genomewide linkage maps of the stickleback. Now we are using these maps to study the genetic changes that occurred in the lakes.

"Each lake is its own little independent evolutionary experiment, and many of these traits have evolved multiple times in different lakes. So we'll be able to ask whether there is only one way to make a fish look a particular way. Are the same genes always involved? That's a very important question in evolutionary biology in general."

In the long term, Kingsley would like to know "what kinds of mutations occur, in what genes, to create the new shapes and behaviors that we see in different species.... In the past decade we've learned a lot about the pathways that create skeletal tissue," he says. "In the next decade I think we'll learn how evolution works to design different organisms."

A FISH WITH A "DISCOUNT GENOME"

Sydney Brenner, who pioneered the genetic analysis of *C. elegans*, looks with pride at the discoveries his former students and others have made—

and are still making—with the little worm. He is happy that its entire genome has been sequenced. Personally, though, he has moved on to another organism: the pufferfish, or Fugu (*Fugu rubripes*), a rather ungainly, spotted fish that puffs up into a ball when it feels threatened.

Fugu is considered a great delicacy in Japan, where it is prized both for its tingling taste and its whiff of danger. Unless properly prepared (bred without certain bacteria or totally cleared of bacterial toxins), Fugu can rapidly paralyze and kill humans. In fact, hundreds of people have died of Fugu poisoning. But Brenner values Fugu for an entirely different reason: its exceptionally compact and gene-rich DNA.

"I call it the 'discount genome,'" says Brenner, who does much of his research at the Molecular Sciences Institute in Berkeley, California. "Fugu's genome is only 400 megabases, about one-eighth the size of the human genome, yet it has roughly the same number of genes as the human, as we have proved."

This gene density makes Fugu DNA much easier to work with than human DNA, Brenner explains. It is difficult to identify human genes in a sea of genomic DNA, even after the DNA has been fully sequenced, since protein-coding genes are usually made up of several pieces separated by "introns" (intervening sequences) and represent less than 5 percent of the total human genome. The vast bulk of the genome consists of stretches of uninterpretable DNA that include many repetitive sequences between genes, as well as in the introns that interrupt the genes and are later spliced out to form messenger RNA. In Fugu, however, there is very little repetitive DNA and the introns are much smaller, yet the coding sequences are the same size. As a result, genes make up 20 percent of Fugu's total DNA. "This means you don't have to sequence all that rubbish," Brenner says. "We can find genes in the fish genome which you can't find in the human, but once you find a gene in the fish, you can look for it in the human."

Even more important, according to Brenner, Fugu will help scientists understand "how genes get switched on and off." Solving this problem will be the task for biology in the next few years, he says—"not just to decipher the mechanism of switching, but to understand the language, what is specified."

There are two ways to view a gene's function, Brenner says. One is to identify the protein that the gene makes and its molecular activity—the nuts-and-bolts approach. The other way is to define "the gene's meaning to the organism." He uses the *ras* gene as

Fugu can rapidly paralyze and kill humans. In fact, hundreds of people have died of Fugu poisoning.

This pufferfish, Fugu, can swell into a ball when it feels threatened. Its value for researchers, Brenner explains, is that "it has no junk DNA. In one-eighth the size of the human genome, it's got all of the genes."

an example. "It's a GTPase, an enzyme that interacts with other things in a signaling pathway. That's all it knows, and that's what you can read from the protein sequence," he says. "But in *Drosophila, ras* has a meaning in the eye; in *C. elegans,* it has a meaning in the vulva; in humans, it has a meaning in the regulation of cell growth. In yeast it's something different again. In each case, the meaning of the pathway is completely different.

"It's nice to know there are these mechanisms and pathways and they're invariant," he adds. "Well, they jolly well should be invariant! It's their control, their meaning, that changes."

In trying to decipher what a gene means for an organism, biologists need to be wary, Brenner warns. A gene may be expressed because its particular product is needed in a particular cell at a particular time—or it may be turned on simply because the organism can't be bothered to turn it off.

"Biology actually has three values: yes, no, and don't care," he declares. "That's what evolution generates.... You can, for instance, have the organism's genes turned on by a general mechanism so they're all on, and then you switch them off in various places. Or you can have them all off, and then switch them on in various places. Okay? Now, for things that are common to many cells, it's best to do it the first way—that is, to say no to some genes. For genes whose products are present in only a few cells, it's best to do it the second way, which is to say yes only to the ones you need there.

> "We could understand why we are different from chimpanzees...."

"The trouble arises when you come to more complicated combinations of these, and now you have to specify combinatorial functions. The solution may be to say, 'Look, I'll have these genes on in these cells; for the others, I don't care whether they are on or off.' This minimizes the regulation." Unless scientists understand which strategy is used, he says, they risk jumping to false conclusions when they see that a gene is switched on.

Brenner believes it is important to identify the "control words" that carry out such strategies, and he has started to do so. "Essentially, we've let nature do the experiments for us," he says. "We've taken two lineages that separated 450 million years ago and focused on certain systems they have in common, such as the immune system or the neuroendocrine system. We wanted to test the hypothesis that not only are the genes in these systems conserved, their regulation is conserved as well.

"If I had two human lineages that separated 450 million years ago, I'd have no problem in finding out, because during that time anything that was nonessential would have been randomized by mutation, and only that which is essential would be retained. Of course I don't have two human lineages that separated then, but I've got humans and I've got fish; and by focusing on common, highly conserved functions, you can begin to find the control words."

He has already identified a few of these words, he says. Working with a region of Fugu DNA that contains a homolog of the human gene for oxytocin (a brain hormone), Brenner and colleagues at the National University of Singapore inserted this bit of Fugu DNA into a rat's genome. (They chose the rat because there are excellent maps of its brain structures, Brenner explains.) "We found that the fish oxytocin was specifically expressed in the rat, in the same cells as the rat oxytocin," he reports. "And it also responded to the same signals. So, as far as the rat is concerned, it's reading the information there in exactly the same way.

"This means that those two pieces of DNA from fish and rat, with their many control sequences, have the same value. Therefore, we could sequence these two pieces and see what is similar, or identical, between them. And we found little fragments of DNA, 14 base pairs, that were totally identical. These are the control words."

Seeking such control signals has been made easier by scientists' growing ability to mix and match pieces of different organisms, Brenner says. "We're not committed, as we were when I started working with *C. elegans*, to having one organism in which you would do everything," he points out. "Today we can range far and wide. We can get the genes from one source and the cells or organism from a different source.

"For example, we've just done an experiment with Mark Kirschner of the Harvard Medical School, looking at the expression of pufferfish genes in frogs. I'm very excited. It might be possible to read the entire genome of the pufferfish by putting it into frogs' eggs."

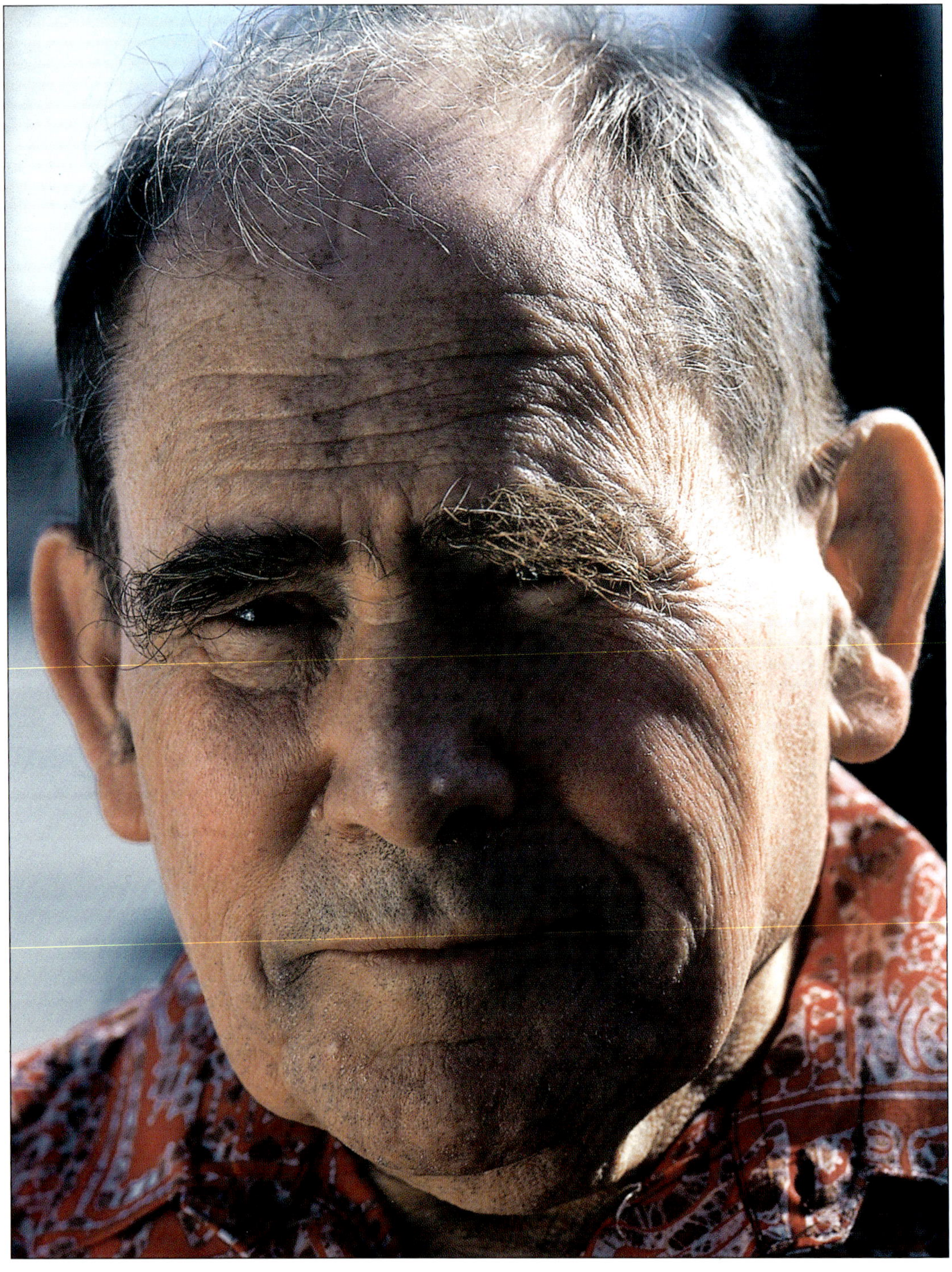

Sydney Brenner keeps opening new fields of research in his drive to learn how genes generate a functioning organism. He recently received a special award from the Lasker Foundation for his "50 years of brilliant creativity in biomedical science."

Ultimately, Brenner would like to define all the control words in human DNA—the commands that shape our physiology. "If we could understand the controls, we could start the real analytic work," he says. "We could reinvent physiology. And we could understand why we are different from chimpanzees or from any other of the great apes. That, I think, will be the most exciting thing we can do." ●

Editor .... Maya Pines
Copy Editor .... Dean Trackman
Production Editor .... Patricia A. Foster
Permissions Editor .... Kimberly A. Blanchard

Art Direction and Design: RCW Communication Design Inc.

Editorial Director: Robert A. Potter

Cover: Background-*Science*, vol. 287, 24 March 2000 © 2000 AAAS (adapted by RCW Communication Design Inc); researcher photo-Kay Chernush; mouse-Robert E. Hammer and Masashi Yanagisawa; yeast-© Jeremy Burgess/Photo Researchers, Inc.; sticklebacks-© OSF/Animals Animals; worm-Erik Jorgensen; fly-John Kachik

2 Jijie Chai, Chunying Du, Jia-Wei Wu, Saw Kyin, Xiaodong Wang & Yigong Shi, *Nature*, vol. 406, cover, 24 August 2000 © 2000 Macmillan Magazines Ltd.
7 Kay Chernush
8 Karen Barnes, Wood Ronsaville Harlin, Inc.
9 © Jeremy Burgess/Photo Researchers, Inc.
11 Audrey P. Gasch
12 Patrick O. Brown and David Botstein, *Nature Genetics Supplement*, vol. 21, p. 34, fig. 1, January 1999 © 1999 Nature America Inc. (adapted by Eade Creative Services, Inc./George Eade illustrator)
13 Karen Barnes, Wood Ronsaville Harlin, Inc.
14 Shelley Chu (adapted by Eade Creative Services, Inc./George Eade illustrator)
15 Kay Chernush
17 William K. Geiger
18-23 Kay Chernush
24 William K. Geiger
24-25 Sergey Vorobiev
26 John C. Matese
27 Adapted by RCW Communication Design, Inc.
29 Eric Schabtach and Ira Herskowitz
30 Eade Creative Services, Inc./George Eade illustrator
32-52 Fly illustrations-John Kachik
33 Paul Fetters
35 E.B. Lewis
37 William McGinnis (adapted by John Kachik)
40-41 Kay Chernush
44 Barbara Ries
46 From "Basal Cell Carcinoma, the Most Common Cancer," © 1986 by the Skin Cancer Foundation
47 Matthew Scott
49, 50 Kay Chernush
53 Ulrike Heberlein (adapted by Eade Creative Services, Inc./George Eade illustrator)
54-55 Erik Jorgensen
57 Paul Fetters
58-59 Lineage map-John Sulston (adapted by Mary E. Challinor); worm-H. Robert Horvitz (adapted by Wood Ronsaville Harlin, Inc.)
60 Lennart Nilsson, *A Child Is Born*, p. 117, 1990 © 1990 by Dell Publishing Co.
62 Brad Hersh
65 Kay Chernush
67 Mario de Bono and Cornelia I. Bargmann, *Cell*, vol. 94, p. 684, fig. 4, September 1998 © 1998 by Cell Press
68 Kay Chernush
69 Katharine S. Liu and Paul W. Sternberg, *Neuron*, vol. 14, No. 1, cover, January 1995 © 1995 by Cell Press; inset-Maureen Barr
70-71 Stuart Kim
73-75 Kay Chernush
77 From *Springfield Sunday Republican*, October 5, 1913 (adapted by RCW Communication Design, Inc.)
78 Paul Fetters
80, 81 Kay Chernush
82 Scott Dingman
85 Christopher Denney
87 Robert E. Hammer and Masashi Yanagisawa
89 Kay Chernush
90 © Mark Smith/Photo Researchers, Inc.
91 Larry Maglott for Children's Hospital
93 Kay Chernush
95 © OSF/Animals Animals
97 B. Venkatesh
99 Kay Chernush

Printing: St Ives Inc.

03-01-002-150K

The Unity of Life

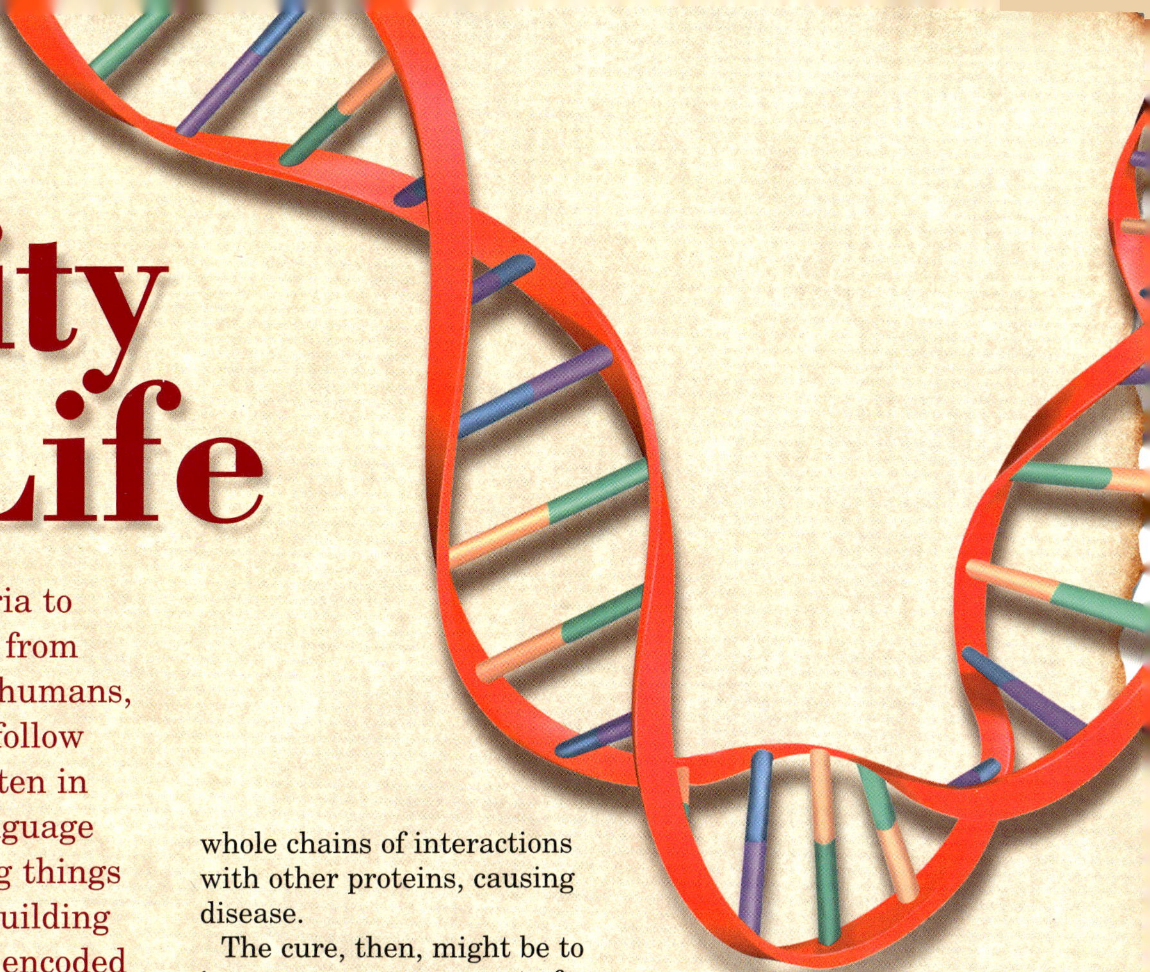

From bacteria to elephants, from flowers to humans, all living things follow instructions written in the universal language of DNA. All living things contain similar building blocks—proteins encoded by DNA. And all diseases can be traced back to malfunctioning genes or proteins.

To speed up the search for better treatments, some scientists now want to move on from genomics, the study of all the genes in an organism's cells, to the next step—proteomics, the study of all the proteins specified by these genes and how the proteins interact.

Proteins are the body's beams and rafters, movers and engineers, as well as message givers and infection fighters. But proteins don't act alone—they bind to other proteins, affecting them. So when a mutant gene produces a defective protein, it can mess up whole chains of interactions with other proteins, causing disease.

The cure, then, might be to interrupt—or compensate for—some of the faulty interactions. But first these need to be precisely identified. This is where model organisms such as yeast and flies are proving particularly useful.

A decade ago, Stanley Fields, an HHMI investigator at the University of Washington, Seattle, devised an ingenious way to identify pairs of proteins that physically interact with one another. Now he and his collaborators are using this "two-hybrid" system to explore the protein interactions in yeast. The scientists recently identified 957 interactions involving 1,004 yeast proteins. Similar interactions are very likely to exist between the corresponding proteins in humans.

Meanwhile Stuart Schreiber and his colleagues at Harvard University have adapted Patrick Brown's microarrays technique—originally devised for DNA—for use with proteins, enabling them to study more than 10,000 proteins simultaneously. In this way, they detected large numbers of previously unknown protein interactions. They also screened hundreds of small molecules to see which ones would interact with the proteins in the microarrays.

Both of these approaches are providing new leads for a wide array of potential new drugs, as well as laying the groundwork for a far more precise medical science.

BLOOD

Leukemia
chronic myelogenous leukemia,
a cancer of the blood (*ABL1* gene)

Bruton Agammaglobulinemia
lack of mature B cells (*BTK* gene)

G6PD Deficiency
drug- and stress-induced rupture of
red blood cells (*G6PD* gene)

LUNG

Cystic Fibrosis
progressive disease of lungs and pancreas
(*CFTR* gene)

Lung Cancer
caused by defects in *P53* gene,
which can also cause cancer of esophagus,
colon, brain, lung, breast, and skin (*P53* gene)

LIVER

Wilson Disease
buildup of copper in cells, causing
liver disease and other symptoms (*ATP7B* gene)

COLON

Hereditary Nonpolyposis Cancer
(*MSH2* gene)

Adenomatous Polyposis
polyps that become malignant (*APC* gene)

BONES

Multiple Exostoses
ossification at tips of femur, pelvis, or ribs (*EXT1* gene)

PROSTATE

Advanced Cancer of the Prostate
caused by mutations in *PTEN* gene, which can also cause
cancer of brain, endometrium, and breast
(*PTEN* gene)

HUM

DISEAS

That Are Found in

Flies don't get kidney disease, and worms don't get heart disease, yet many of the human genes that are faulty in these and other human disorders have parallel genes in model organisms, where they can be studied more easily.

After the fly's genome was sequenced in March 2000, a team of scientists found that 61 percent of the human genes known to be mutated in 289 human diseases have close equivalents in flies. Many of these genes in yeast.

The extent of similarity genes and genes that have yeast is shown below. Gen similarity are colored dark And a lower degree of sim

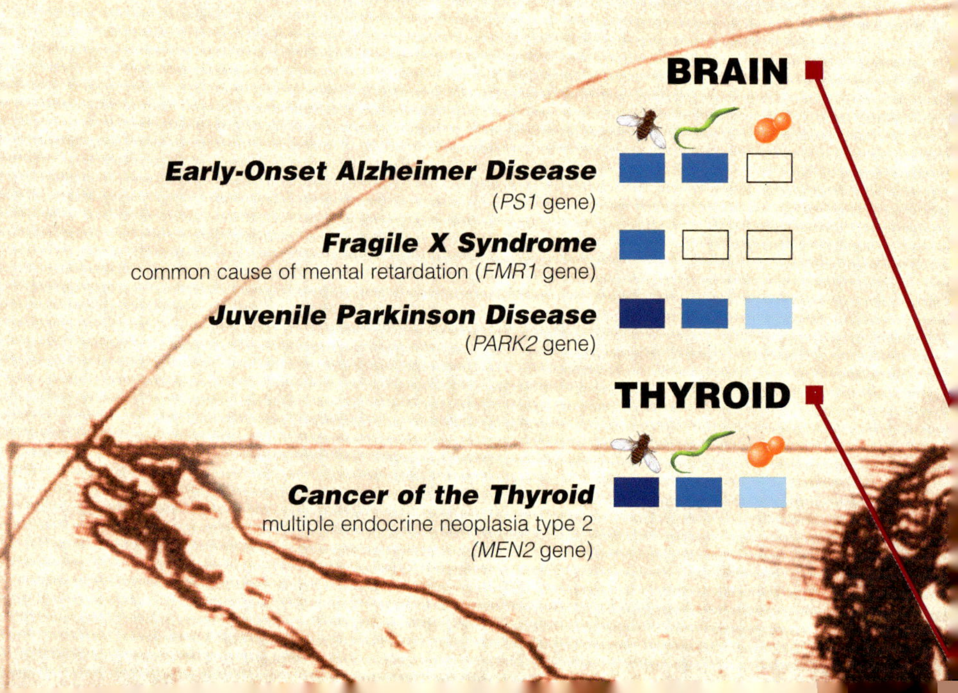

The four model organisms shown here—yeast, worm, fly, and mouse—share vast numbers of genes, proteins, and even genetic pathways with humans.

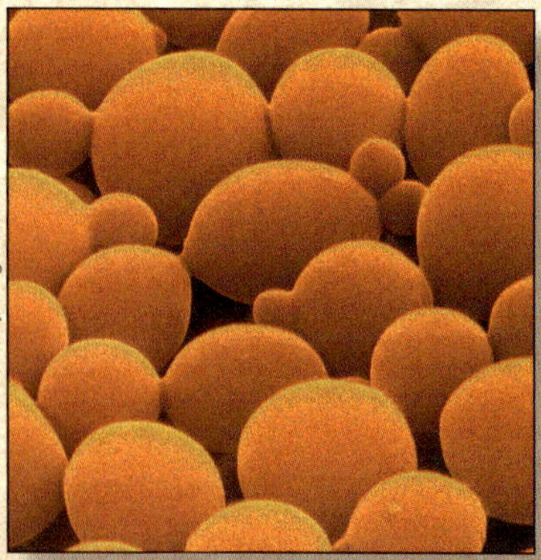

© Jeremy Burgess/Photo Researchers, Inc.

Yeast

(Saccharomyces cerevisiae)

- A single, free-living cell, only 3 microns in diameter (4,000 of them lined up would measure an inch).
- Reproduces by budding and doubles every 90 minutes.
- Its genome was sequenced in 1996.
- 12 million base pairs of DNA.
- 6,000 genes, of which at least 31 percent have human equivalents.

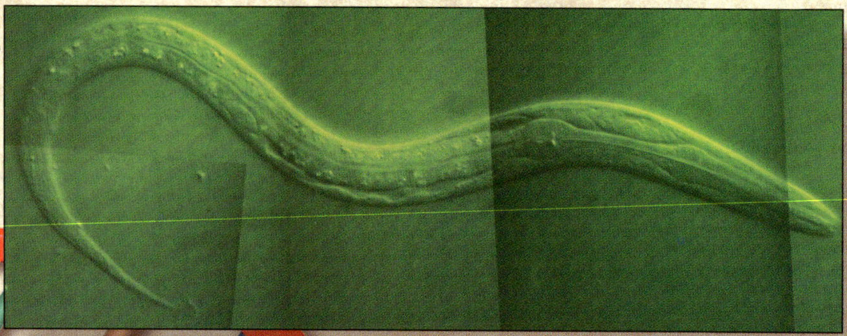

Erik Jorgensen

Worm

(Caenorhabditis elegans)

- A multicellular animal, 1 millimeter (0.04 inch) long.
- Lifespan: 2–3 weeks. A new generation every 3 days.
- Its genome was sequenced in 1998.
- 99 million base pairs of DNA.
- 19,099 genes, of which 40 percent have human equivalents.

The Unity of Life

Robert E. Hammer and Masashi Yanagisawa

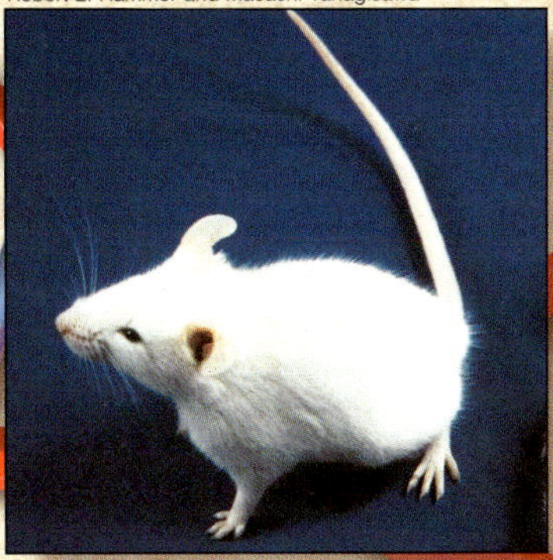

Mouse

(Mus musculus)

- Our closest relative among model organisms, 170 millimeters (6.6 inches) long.
- Lifespan: 2 years. A new generation every 9 weeks.
- Its genome sequence is expected in 2001.
- An estimated 3 billion base pairs of DNA (as in human).
- An estimated 40,000 genes (as in human).
- Almost every human gene has a counterpart in the mouse, and some blocks of sequenced mouse DNA are proving impossible to tell apart from the human versions.

Donny Bliss

John Kachik

Fly

(Drosophila melanogaster)

- A multicellular animal with complex behavior, 4 millimeters (0.16 inch) long.
- Lifespan: 2–3 months. A new generation every 10 days.
- Its genome was sequenced in March 2000.
- 165 million base pairs of DNA.
- 13,600 genes, of which about 50 percent have human equivalents.

Human

(Homo sapiens)

- 5–6 feet tall.
- Lifespan: About 40 years in developing nations, 60–70 years in the United States and other industrial nations. A new generation every 20–25 years.
- The human genome was sequenced (preliminary draft) in June 2000.
- 3 billion base pairs of DNA.
- An estimated 40,000 genes.

Despite their obvious differences in size and way of life, all these model organisms make proteins that carry out the same core functions as in humans: telling the organism when and how to grow, reproduce, fight off stresses, and eventually die.

MAN

GENES

lies, Worms, & Yeast

e parallels in worms and even

n some of the human disease quenced in flies, worms, and the highest degree of sequence he next group is medium blue. shown in pale blue. Blank boxes indicate extremely low or no similarity between the genes.

"Some of the human disease genes can be found in yeast—they are very ancient genes," says Gerald Rubin, the report's lead author.* "Some can be found only in worms and flies, and some only in flies. Of course the closest similarity would be with the mouse. Any gene on this list would have a parallel in the mouse."

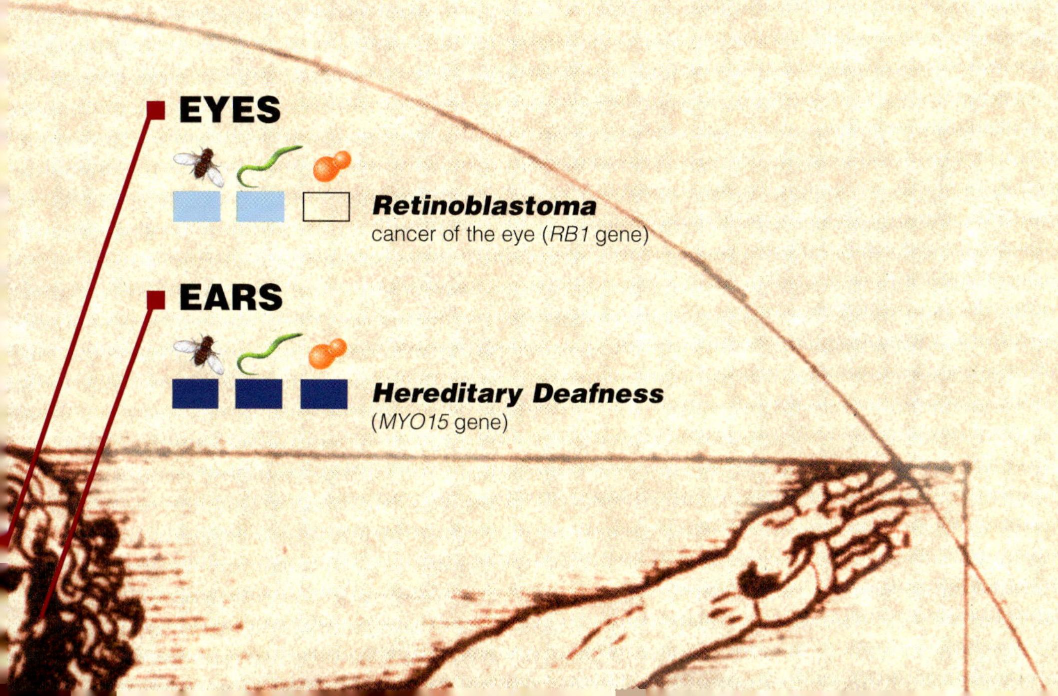

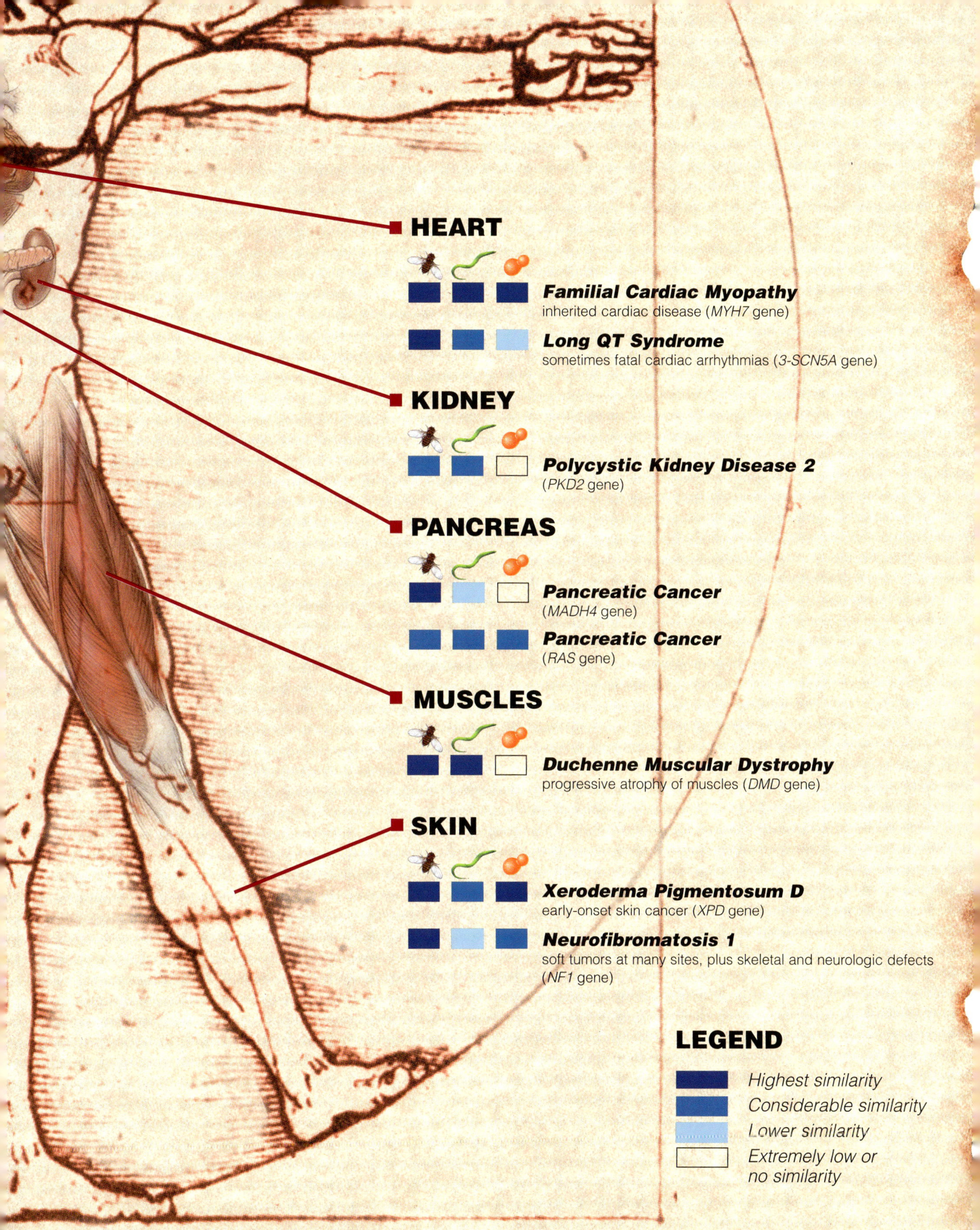

OURCE: *G.M. Rubin et al., "Comparative Genomics of the Eukaryotes,"* Science *287 (March 24, 2000): 2210–11, fig. 1.*

Communication Design Inc.; interior anatomy and yeast illustrations-Donny Bliss; fly-John Kachik; worm-Erik Jorgensen

A Dramatic Rescue

An experiment involving the eyes of flies offers a promising example of how research on model organisms may lead to new types of treatment for human diseases—even for such seemingly hopeless ones as Alzheimer and others that result from damaged or dying brain cells.

Flies were chosen for this experiment because, as Nancy Bonini of the University of Pennsylvania explains, "late-onset, progressive diseases such as Alzheimer or Parkinson can take months or years to develop in mouse models, but in flies they take only 10 days."

Bonini focused on a group of inherited brain diseases that are caused by a peculiarly large number of repeats of the DNA triplet CAG, which codes for the amino acid glutamine. These diseases include Huntington, which killed the folksinger Woody Guthrie; spinocerebellar ataxia type 3 (SCA3, also known as Machado-Joseph disease); and at least six other ailments that typically begin in adulthood. In each case, people whose gene has fewer than 35 CAG repeats stay healthy, while those whose gene typically contains more than 40 repeats become ill. Apparently the proteins that result from this excess of glutamine cannot fold into their proper shapes. As a result, they become toxic to neurons and slowly kill them off.

The experiment showed how such proteins can be helped to recover their normal shapes. Bonini and her colleagues at the University of Pennsylvania and the University of Iowa knew that a special class of proteins, called chaperones, has the job of guiding other proteins toward their proper structure. Would an extra supply of these chaperones prod the proteins into line sufficiently to prevent them from becoming toxic to neurons?

To find out, the researchers first made transgenic flies that contained the human *SCA3* gene—and the flies' retinas rapidly degenerated. Next they made flies with both the *SCA3* gene and a gene for a human chaperone protein, HSP70. This led to a dramatic rescue: The chaperone suppressed the disease and the eye looked normal again (see photo below, right).

"Many different human neurodegenerative diseases appear due to toxic protein structure," Bonini says, "so these findings may apply to a large number of disorders."

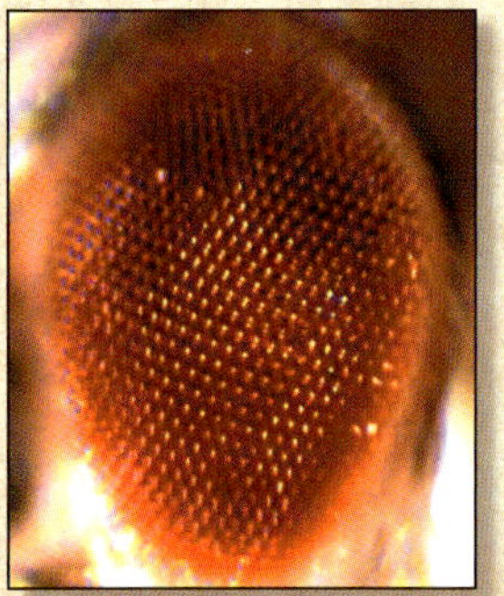

A normal *Drosophila* eye with regular structure and full pigment. The cells in each of the eye's 800 identical units are arranged in a precise pattern.

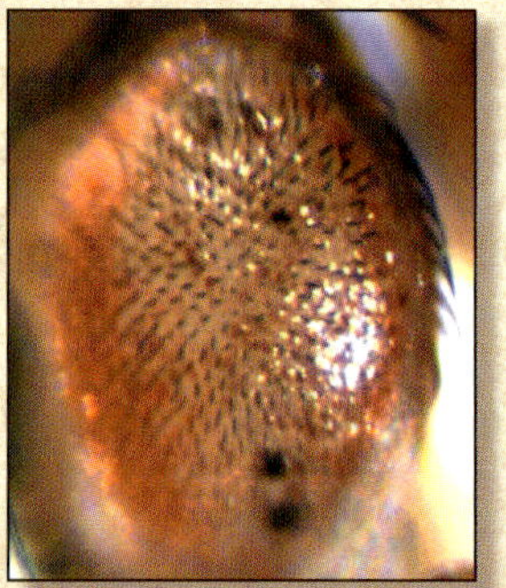

Many of this transgenic fly's retinal neurons are sick and its retina has degenerated because of toxic proteins made by an inserted human *SCA3* gene.

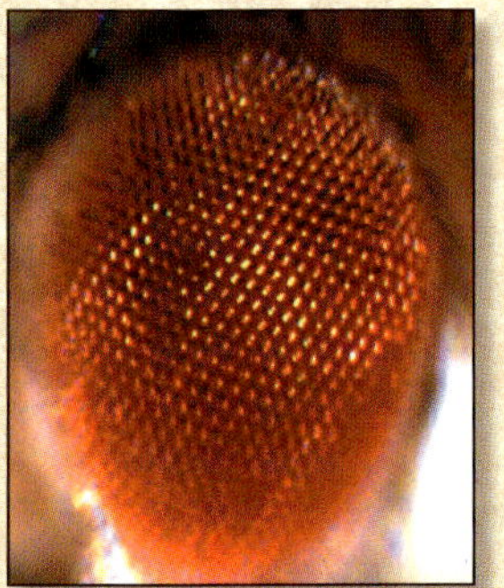

After treatment—an extra supply of chaperone proteins—the fly's eye appears normal externally and its internal structure is largely restored.

The Unity of Life